T0133276

V&Runipress

Brigitte Elisabeth Slaats

Investigations on the efficacy of encapsulation of the endoparasitic fungus *Hirsutella rhossiliensis* for control of plant-parasitic nematodes

V&R unipress

Bibliografische Information der Deutschen Nationalbibliothek

Die Deutsche Nationalbibliothek verzeichnet diese Publikation in der Deutschen Nationalbibliografie; detaillierte bibliografische Daten sind im Internet über http://dnb.d-nb.de abrufbar.

ISBN 978-3-89971-479-1

© 2008, V&R unipress in Göttingen / www.vr-unipress.de

Dedicated to my family

Table of Contents

Abstract

Investigations on the efficacy of encapsulation of the endoparasitic fungus *Hirsutella rhossiliensis* for control of plant-parasitic nematodes

Hirsutella rhossiliensis is an endoparasitic fungus that parasitizes a multitude of plant-parasitic nematodes, which are found in agricultural soils world-wide. For a commercial exploitation of *H. rhossiliensis* as a biological control agent, a proper formulation is required. Until now, a suitable formulation has not been found. In a joint project between BBA, FAL, KWS SAAT AG and BIOCARE GmbH, renewable resources were investigated as encapsulation material for *H. rhossiliensis*. The control potential of various types of encapsulation was investigated against the sugar beet cyst nematode *Heterodera schachtii* and the root-knot nematode *Meloidogyne incognita*. Prior to the encapsulation studies, laboratory assays showed that the *H. rhossiliensis* isolate BBA was pathogenic towards different sedentary and migratory endoparasitic nematode species. The side-effect on non-target entomogenous nematodes was minimal. Furthermore, the use of *H. rhossiliensis* in integrated pest management systems was investigated by testing the side-effects of synthetic plant protection agents on this fungus. Regarding integration of *H. rhossiliensis* application with common plant protection agents in sugar beet production and greenhouse production of tomato and cucumber, most pesticides do not seem to harm the fungus and therefore can be applied together. The encapsulation of *H. rhossiliensis* in derivates of Guargum MG, Guargum MF, pectine and alginate was tested in various greenhouse trials against *H. schachtii* on sugar beet and *M. incognita* on cucumber. Results showed that nematode invasion into the roots of seedlings could be reduced if the capsules containing *H. rhossiliensis* were applied to heat-treated soil. If fungal capsules were applied to non-treated soil, capsules were colonized by soil microorganisms and growth of *H. rhossiliensis* was inhibited. Attempts to reduce microbial colonization of the capsule by changing the capsules pH to be unattractive for soil microorganisms failed to promote growth of *H. rhossiliensis*. Existing difficulties in the growth of the fungus in field soil and the connecting competition towards other soil microorganisms could not be solved to satisfaction with this technique. Alternatively, liquid and solid formulations of *H. rhossiliensis* were tested. Liquid formulations led to a reduction of root penetration by 50% in heat-treated soil. But neither liquid nor solid formulations achieved a reduction of the final population density in a greenhouse trial conducted with *H. schachtii* on sugar beet. Due to its weak competitiveness, *H. rhossiliensis* is considered a poor candidate for biological control of plant-parasitic nematodes.

Zusammenfassung

Untersuchungen zur Wirksamkeit von Verkapselungen des endoparasitären Pilzes *Hirsutella rhossliensis* zur Bekämpfung von pflanzenparasitären Nematoden

Hirsutella rhossiliensis ist ein endoparasitärer Pilz, der weltweit in landwirtschaftlichen Böden vorkommt und eine Vielzahl von pflanzenparasitären Nematoden befällt. Die kommerzielle Nutzung des Pilzes setzt eine entsprechende Formulierung voraus. In einem Gemeinschaftsprojekt der BBA mit der FAL, KWS SAAT AG und BIOCARE wurden nachwachsende Rohstoffe als Verkapselungsmaterial für *H. rhossiliensis* eingesetzt. Ihre Eignung wurde gegen den Zuckerrübenzystennematoden *Heterodera schachtii* und den Wurzelgallennematoden *Meloidogyne incognita* untersucht. Vor der Verkapselung des BBA-isolates *H. rhossiliensis* zeigten Laborversuche, dass das Isolat gegenüber verschiedene, sedentäre und wandernde endoparasitäre Nematoden pathogen war. Die Nebenwirkung auf Nicht-Zielorganismen war gering. Um seine Verwendung in einem integrierten Bekämpfungssystem zu überprüfen, wurden die Nebenwirkungen von ausgewählten Pflanzenschutzmitteln auf das Wachstum des Pilzes untersucht. In Hinblick auf die Integration einer Applikation von *H. rhossiliensis* mit herkömmlichen Pflanzenschutzmitteln im Zuckerrübenanbau und in den Gewächshauskulturen Tomate und Gurke scheinen die meisten Pestizide den Pilz nicht zu schädigen und könnten somit gemeinsam appliziert werden. Die Verkapselung von *H. rhossiliensis* in Derivaten von Guargum MG, Guargum MF, Pektin und Alginat wurde in Gewächshausversuchen gegen *H. schachtii* an Zuckerrüben und *M. incognita* an Gurken untersucht. Die Applikation von Pilzkapseln in gedämpfter Erde führte zu einer Befallsreduktion. Bei Applikation in ungedämpfter Erde wurden die Pilzkapseln jedoch von Fremdorganismen befallen, so dass *H. rhossiliensis* nicht aus den Kapseln auswuchs. Eine Veränderung des pH-Wertes der Kapseloberfläche, um deren Attraktivität für bodenbürtige Mikroorganismen zu reduzieren, führte nicht zu einem verbesserten Auswachsen des Pilzes. Bestehende Schwierigkeiten beim Auswachsen des Pilzes im Boden und der damit verbundenen Konkurrenzfähigkeit gegenüber anderen Bodenmikroorganismen konnten mit dieser Technik bisher aber nicht zufriedenstellend gelöst werden. Alternativ wurden feste und flüssige Formulierungen von *H. rhossiliensis* in Gewächshausversuchen untersucht. Obwohl eine Reduktion des Wurzelbefalls mit *H. schachtii* in gedämpfter Erde festgestellt wurde, konnte die Besatzdichte von *H. schachtii* im Boden nicht nachhaltig gesenkt werden. Insbesondere die schwache Konkurrenzkraft von *H. rhossiliensis* schränkt eine Nutzung dieses Pilzes in der biologischen Bekämpfung ein.

Abbreviation Index

ALG	alginate derivate
approx.	approximate(ly)
BBA	Federal Biological Research Centre for Agriculture and Forestry
©	Copyright
°C	degrees Celsius
cfu	colony forming units
cm	centimetre
CON	Non-treated control
d	day (time)
H.r.	*Hirsutella rhossiliensis*
e+j	eggs and juveniles
FAL	Federal Agricultural Research Centre
e. g.	exempli gratia – »for example«
g	gram
h	hour (time)
h	height
i.e.	id est – »that is«
kg	kilogram
l	liter
j2	infective second stage juvenile of *H. schachtii* or *M. incognita*
MF	Guargum MF = guargum derivate MF
MG	Guargum MG = guargum derivate MG
mg	milligram
min	minute (time)
ml	milliliter
mm	millimeter
µl	microliter
µm	micrometer
n	number of observations/replicates (statistics)
no.	number
PA2/PA5	pectine derivates
PDA	potato dextrose agar
Pf	final population density
pH	hydrogen-ion activity, negative logarithm of (hydrogen-ion exponent); a standard used to measure a liquid's acidity or alkalinity on a scale of 0 to 14
Pi	initial population density
rpm	revolutions per minute
SD	standard error (of estimate mean value)
sp.	species
spp.	plural abbreviation, refers to all individual species within a genus
WA	water agar-agar

w/w	weight by weight
Ø	diameter
>	bigger than
<	smaller than
®	Registered symbol

List of Chemicals

Chemicals/Materials	Deliverer/Producer
Agar-Agar (WA)	AppliChem GmbH, Darmstadt
Baker's yeast (dried)	Dr. August Oetker, Nahrungsmittel KG, Bielefeld
Citric acid monohydrate p.A.	AppliChem GmbH, Darmstadt
D (+) Glucose ($C_6H_{12}O_6$)	AppliChem GmbH, Darmstadt
Di-Potassium hydrogen phosphate (K_2HPO_4)	Merck KGaA, Darmstadt
Ethanol 95%	Waldeck GmbH & Co. Kg, Münster
Fuchsin acid (rubin S)	Merck KGaA, Darmstadt
Glycerin 87% ($C_3H_8O_3$)	AppliChem GmbH, Darmstadt
Hydrochloric acid (HCL)	AppliChem GmbH, Darmstadt
Kaolin	Waldeck GmbH & Co. Kg, Münster
Mercury (II) chloride GR ($HgCl_2$)	Merck KGaA, Darmstadt
Magnesiumsulfate ($MgSO_4$)	Harneberg Landhandel, Münster
MES ($C_{16}H_{13}NO_4*H_2O$) 2-Morpholinoethanesulfonic acid	Merck KGaA, Darmstadt
Sodiumhypochlorite (5.25% NaOCl) »Danchlorix«	Colgate-Palmolive GmbH, Hamburg
Penicillin G ($C_{16}H_{17}N_2O_4SK$)	Sigma-Aldrich Chemie, Steinheim
Potato dextrose agar (PDA)	Merck KGaA, Darmstadt
Streptomycin sulphate ($C_{12}H_{39}N_7O_{12}*1.5\ H_2SO_4$)	Serva Heidelberg, Boehringer Ingelheim Bioproducts Partnership
Sodium hydroxide (NaOH)	Riedel-de Häen AG, Seelze-Hannover
Tri-Sodium citrate-2-hydrate ($C_6H_5Na_3O_7*2H_2O$)	Riedel-de Häen AG, Seelze-Hannover
Vanillic acid ($C_8H_8O_4$)	Merck Schuchardt OHG, Hohenbrunn
Yeast extract granulated	Merck KGaA, Darmstadt
Zinc chloride ($ZnCl_2$)	Riedel-de Häen AG, Seelze-Hannover

All the companies are situated in Germany.

1 General introduction

1.1 Important plant-parasitic nematode species

Plant-parasitic nematodes are small microscopic roundworms which live in the soil and attack the roots of plants. Among many different types of plant-parasitic nematodes, only species of about 10 different genera cause most of the damage by nematodes in agriculture and horticulture. Among the important species are the sugar beet cyst nematode *Heterodera schachtii* Schmidt 1871 and the root-knot nematode *Meloidogyne incognita* (Kofoid & White 1919) Chitwood 1949, on which this research was focused. The sugar beet cyst nematode was the first plant-parasitic nematode of economic importance observed in 1859 by A. Schmidt in Germany (Franklin, 1951). Twelve years later, in 1871, A. Schmidt named and described it as *Heterodera schachtii*, the sugar beet cyst nematode (Schmidt, 1871). At that time the sugar beet industry had great economic importance in Central Europe and because of the serious losses to sugar beet in the field by this nematode scientists undertook intensive research on its life-cycle and control (Singh and Sitaramaiah, 1994). Until today, in several European countries, *H. schachti* has been a serious pest in sugar beet production. It is estimated that the annual yield loss in EU countries on the world market sugar price level amounts to 90 million Euros ($ 118 million) (Müller, 1999). Its distribution is world-wide, *H. schachtii* reproduces with few exceptions on plant species within the family Chenopodiaceae and Cruciferae (Krall and Krall, 1978). Root galls caused by nematodes first attracted attention in 1855 when they were discribed by Berkeley on cucumbers in an English greenhouse. Since then it has been a cause for concern world-wide as root-knot nematodes of the genus *Meloidogyne* attack a great number of staple, industrial, vegetable and fruit crops and many weed species (Richardson and Grewal, 1993). They parasitize 3000 wild and cultivated plant species. About 80 *Meloidogyne* species are described (Carneiro et al. 2000). Only four of them account for 95% of all root-knot infestations in agricultural lands: *M. incognita*, *M. javanica*, *M. arenaria* and *M. hapla* (Hussey and Janssen, 2002). The economically important species differ in host range and pathogenicity (Orton Williams, 1975).

1.2 Life-Cycle and symptoms of damage

Cyst and root knot nematodes are sedentary endoparasites that belong to the order *Tylenchida*. Like other plant-parasitic nematodes, they have a life-cycle of six stages: egg, four juvenile stages, and adult. The embryo develops inside the egg to become the first-stage juvenile, which molts to the second-stage juvenile. The second-stage juvenile hatches and is considered the »infective« stage of cyst and root-knot nematodes. The juveniles move through the soil, locate a host root and enter them. Both nematode taxa induce specialized feeding structures called »nurse« or »giant« cells.

The juvenile swells, molts rapidly a second and a third time and finally matures into an adult male or female. The female nematode becomes immobile, and the body swells to a round, lemon, kidney, or ovoid form. Adult males elongate within the fourth-stage cuticle and emerge from roots. Mature females of the sedentary endoparasitic nematodes generally produce up to 500 eggs which remain in their bodies (cyst nematodes) or accumulate in gelatinous masses attached to their bodies (root-knot nematodes). The primary difference between both genera of nematodes is that the root-knot female remains white with a soft cuticle and the cyst nematode turns into a hard-cuticled brown cyst. Sedentary endoparasites damage their hosts by redirecting large amounts of assimilate from the plant to the nurse cells to allow nematode development. The altered tissues of the feeding site also disrupt the vascular system. The most obvious foliar symptom of *H. schachtii* to sugar beet is the premature wilting of heavily infested plants, often in patches corresponding to the most heavily infested areas in the field. Heavily infested roots stimulate the initiation of new lateral roots opposite of the invasion site, with the result that heavily infested roots soon develop a characteristic bearded appearance (Glovatskaya, 1971, Cooke, 1984). Roots infested by *Meloidogyne* species are frequently stunted, abnormally galled, sometimes massively. The lower foliage may be pale in colour; rapid wilting can occur in periods of sunshine. Infested plants are generally unhealthy, but with favourable growing conditions and an abundance of water, some may tolerate attack, especially if part of their root system has avoided attack (Hussey et al. 1969).

1.3 Control measures

Nematology experienced rapid growth in the late nineteenth and early twentieth century. Due to the serious losses of sugar beet in central Europe by the cyst nematode *H. schachtii*, scientists undertook intensive research on its life-cycle and control. In 1881, Julius Kühn, in Germany, demonstrated the first chemical treatment for nematode control by applying carbon disulphide to soil against the sugar beet cyst nematode. He also conducted studies on trap crops and crop rotations to break the life-cycle of the nematode. Crop rotation proved to be the most effective and economic control method and still remains the oldest and most widely used field control measure for nematodes today (Singh and Sitaramaiah, 1994). Crop rotation, however, is only effective for species with a restricted host range. Furthermore, intensification of cropping has been possible through the use of chemical nematicides and the development of cultivars with resistance to nematode multiplication. Growers rarely rely on one method of control and several measures are integrated into a management strategy. Even so, populations able to overcome resistance in plants are now common and the use of nematicides has been associated with environmental contamination and hazards to health. Nematicides include some of the most toxic products used in agriculture and several have been withdrawn from the market (Burrows et al. 1994). Industries must present effective, alternative methods of plant-parasitic nematode control. However, as costs for the production of new products are very high due to immense costs for toxicity testing and environmental impact testing, many chemical manufacturers have given up research and development of new nematicidal compounds. New approaches to nematode control are urgently needed for the development and commercialization of environmentally safe, cost-effective and reliable nematode control products that meet the demands of the grower and consumer such as an effective implementation of biocontrol agents based on fungal antagonists of nematodes.

1.4 Fungal antagonists

Natural enemies of nematodes include fungi, bacteria, viruses and some animals such as insects, mites and nematodes. Although fungal antagonists of nematodes were already detected in the nineteenth century, nematophagous fungi were only considered as potential biological control agents for plant-parasitic nematodes since Zopf (1988) observed that the nematode-trapping fungus, *Arthrobotrys oligospora*, was able to capture and colonize motile nematodes (Kerry and Jaffee, 1997). There are more than 160 fungus species that live on nematodes partially or entirely (Dowe, 1987). Nematophagous fungi include trapping fungi that form special devices to capture and kill nematodes, endoparasites of vermiform nematodes and fungi colonizing eggs and females of sedentary endoparasitic nematodes (Chen, 2004). In general, nematophagous fungi can be divided into two major groups: nematode-trapping fungi and endoparasitic fungi. Examples of nematode trapping fungi are *Dactylaria candida*, *Arthrobotrys oligospora* and *A. superba* (Drechsler, 1937; Shepherd, 1955; Barron, 1975). Nematode-trapping fungi are often referred to as predators as they are saprophytic and develop their mycelium predominantly outside their host. Their prey is trapped on vegetative hyphae or in special traps; only mobile nematode stages can be caught (Dowe, 1987). Nematode species belonging to endoparasitic nematodes are for instance: *Drechmeria coniospora*, *Paecilomyces lilacinus*, *Pochonia chlamydosporium* and *Hirsutella rhossiliensis* (Dijksterhuis et al., 1990; Cabanillas et al., 1988; Kerry and Jaffee, 1997 and Minter and Brady, 1980). The life-cycle of the endoparasite takes place inside the host except for the development and spreading of the spores. Any nematode stage can be attacked depending on the fungal species. Infection occurs in most cases through the cuticle and is therefore attractive for biological control. Fungal antagonists of nematodes continuously destroy nematodes in virtually all soils. Adding fungus inoculum to soil is a direct and rapid approach to nematode control (De Leij et al. 1993, Jaffee and Muldoon, 1995; Meyer et al. 1997; Stirling and Smith, 1998). Although biocontrol microbes often are not thought of as acceptable alternatives for pesticides due to the lack of broad spectrum activity, inconsistent performance, and slower action when compared with pesticides, there have been commercial successes for management of nematodes. The available commercial products registered for biocontrol include formulations containing *Pochonia chlamydosporium and Paecilomyces lilacinus.*

1.5 *Hirsutella rhossiliensis*

Among the most studied biocontrol agents of plant-parasitic nematodes is the endoparasitic fungus *Hirsutella rhossiliensis* Minter & Brady (Minter and Brady,

1980; Sturhan and Schneider, 1980; Kerry and Jaffee, 1997). It is a hyphomycetes with simple erect phialides which are swollen at the base and taper towards the apex. The fungus occurs naturally in agricultural fields and parasitizes vermiform stages of many plant-parasitic nematodes (Sturhan and Schneider, 1980; Jaffee and Muldoon, 1989). *H. rhossiliensis* produces adhesive conidia that attach to the cuticles of nematodes. Only the conidia that are still attached on the conidiogenous cells are infectious (McInnis & Jaffee, 1989) and one conidium is generally sufficient to infect a nematode (Cayrol & Frankowski, 1986) (Figure 1).

Figure 1: Adhesive conidium of Hirsutella rhossiliensis *on conidiophore. Photo by P. Timper.*

Once the conidium is attached to a nematode, it germinates. The germ tube directly penetrates the host cuticle, and assimilative hyphae grow through and consume the nematode. After colonizing and killing the host, the fungus grows from the cadaver and produces a new cohort of conidia (Sturhan and Schneider, 1980; Jaffee and Zehr, 1982; Jaffee, 1992). This fungus is an attractive candidate as a potential biological control agent for it effectively parasitizes nematodes in natural soil. Suppression of *H. schachtii* populations on fodder radish in Germany was reported by Müller, 1982. In a field in Minnesota, USA, numbers of the soybean cyst nematode *Heterodera glycines*, remained small in soybean monoculture over 27 years, which could be attributed to high rates of parasitism by *H. rhossiliensis*; *H. rhossiliensis* parasitized 11-53% of J2 of *H. glycines* during the soybean growing seasons (Chen, 1997). Furthermore this fungus has a broad host spectrum. In greenhouse or laboratory assays, the fungus suppressed *Globodera pallida* on potato (Velvis and Kamp, 1996), *H. schachtii* on cabbage (Jaffee and Muldoon, 1989), *Pratylenchus penetrans* on potato (Timper and Brodie, 1994), *Heterodera glycines* on soybean (Liu and Chen, 2001), and *M. incognita* on tomato (Amin, 2000). To achieve biological control of vermiform nematodes via inundative addition of *H. rhossiliensis* to soil, three potential forms of *H. rhossiliensis* inoculum were considered by Jaffee et al. (1996): conidia, colonized host nematodes and assimilative hyphae. Conidia of *H. rhossiliensis* were not useful for infestation of soil because they did not adhere to nematodes unless they produced newly formed conidia *in situ* (McInnis and Jaffee, 1989). In contrast, the colonized host is an effective form of inoculum. When *H. rhossiliensis*-colonized nematodes were added to soil microcosms, substantial proportions of *H. schachtii* were parasitized (Jaffee et al. 1992). Although colonized nematodes can be produced in the laboratory, the procedure is labor intensive and probably unsuitable for commercialization or even for large-scale field experimentation. For inundative addition of *H. rhossiliensis* to soil, assimilative hyphae have been considered as a substitute for the colonized nematode. Assimilative hyphae, which are produced within the host or in shake culture, supported sporulation when added to a variety of soils (Lackey et al. 1992). *H. rhossiliensis* has been studied using various formulations against many species of plant-parasitic nematodes, e. g. solid cultures of the fungus on corn grits against the soybean cyst nematode *H. glycines* (Liu and Chen, 2001; Chen and Liu, 2005), alginate pellets against *H. schachtii* and root knot nematodes (Lackey et al., 1993, 1994; Tedford et al., 1995; Jaffee et al., 1996; Jaffee and Muldoon, 1997; Jaffee, 2000), fungus-colonized nematodes against *H. schachtii* (Jaffee, 2000) and *P. penetrans* (Timper and Brodie, 1994), mycelium fragment suspension against *G. pallida* (Velvis and Kamp, 1996), suspension of liquid culture colonies against *Meloidogyne hapla* (Viaene and Abawi, 2000) and spore suspension against *M. incognita* (Amin, 2000).

In the following studies, we chose encapsulation of assimilative hyphae of *H. rhossiliensis* primarily for control of the sugar beet cyst nematode *H. schachtii* but also for control of the root-knot nematode *M. incognita*. Encapsulation of *H. rhossiliensis* could be more suitable than other formulations because it can easily be developed, produced, dried and stored. Moreover, the end product would be a granule with good handling properties for the grower.

The following research was conducted within the project entitled »Polymers based on renewable resources for biological control of the sugar beet cyst nematode and fungal pathogens of root rot for an ecological sugar beet production«, a cooperation of the Federal Biological Research Centre for Agriculture and Forestry (BBA), Münster, the Federal Agricultural Research Centre (FAL), Braunschweig, KWS SAAT AG (KWS Seed Company), Einbeck and BIOCARE GmbH, Einbeck, Germany. The project was funded by the Agency of Renewable Resources (FNR; Gülzow, Germany; FKZ: 22015501/ 22003602/ 22003702/ 22003802). Our goal was to develop a biocontrol agent by optimizing existing capsule systems containing *H. rhossiliensis* that had already been applied against *H. schachtii* (Lackey et al., 1993, 1994; Tedford et al., 1995; Weißenborn 1995; Isemer, 1996; Jaffee et al., 1996; Patel et al. 1996; Jaffee and Muldoon, 1997; Gutberlet, 2000; Jaffee, 2000; Rose, 2000).

1.6 Objectives

The objectives of the following studies were to:
- Determine the host suitability of *H. rhossiliensis* against other plant-parasitic nematode species and the effect of the fungus on non-target organisms such as entomogenous nematodes.
- Study the compatibility of the fungus towards various synthetic plant control agents.
- Test different materials based on polymers of renewable resources for encapsulation of *H. rhossiliensis*.

- Demonstrate the efficacy of encapsulation of *H. rhossiliensis* against the plant-parasitic nematode species *H. schachtii* and *M. incognita in situ*.
- Study the effect of additives to enhance the biocontrol potential of encapsulated *Hirsutella rhossiliensis* towards *Heterodera schachtii* and *Meloidogyne incognita*.
- Test the efficacy of alternative liquid and solid formulations against *H. schachtii*.

2 General Material and Methods

2.1 Test Plants

2.1.1 Beta vulgaris

The majority of biotests and greenhouse pot trials on sugar beet (*Beta vulgaris*) against *Heterodera schachtii*, were conducted with sugar beet ›Dorena‹ (KWS Saat AG, Einbeck, Germany). Seeds of the nematode susceptible variety ›Dorena‹ were routinely coated with the fungicides Thiram® (Bayer CropScience, Leverkusen, Germany) and Hymexazol® (Sankyo Co., Ltd. Tokyo, Japan) and the insecticides Imidacloprid® (Bayer CropScience AG, Leverkusen, Germany) and Tefluthrin® (Syngenta AG, Basel, Switzerland). In first studies on encapsulation of *H. rhossiliensis* with Guargum MF+PA5 against *H. schachtii* the nematode susceptible sugar beet variety ›Penta‹ was used, which was coated with the fungicides Thiram® and Hymexazol® and the insecticide Imidacloprid®. If two to five-day-old sugar beet seedlings were used in the trials, then sugar beet seeds were sown into germination trays filled with silicate sand and placed in the greenhouse at 20°C ± 3°C and watered daily. Two to five days after the first seedlings germinated, they were removed from the trays. Roots were rinsed in tap water and shortened to a length of about 3.5 cm before seedlings were transplanted into the soil of a container or pot. If seeds were directly sown into a container or pot, seeds were individually placed into a seed hole about 1 cm deep.

2.1.2 Solanum lycopersicum

The tomato cultivar ›UC 82 Davis‹ (ISI Sementi, Fidenza, Italy) was used for trials on tomato (*Solanum lycopersicon*) against the root-knot nematode *Meloidogyne incognita*. It is highly susceptible towards this nematode species. Greenhouse trials on tomato were conducted in 100 ml containers, in which two tomato ›UC 82 Davis‹ seeds were individually sown about 0.5 cm deep. Both seedlings remained in a container until the trial was harvested.

2.1.3 Cucumis sativus

The nematode susceptible cucumber cultivar ›Belcanto F1‹ (Nebelung GmbH & Co., Warendorf, Germany), was used for 100 ml biotests on cucumber (*Cucumis sativus*) against *M. incognita*. Two seeds were individually sown 1 cm deep into the soil of each container. Germinated seedlings remained in the containers until the trial was harvested.

2.2 Nematode inoculum

2.2.1 Preparation of cyst inoculum of Heterodera schachtii

In order to infest soil of experiments with cysts containing eggs and juveniles of *H. schachtii*, a culture of this nematode was established on fodder radish ›Siletina‹ (*Raphanus sativus oleiformis*) (Nebelung GmbH & Co., Warendorf, Germany). For this, fodder radish seeds were sown into 12 cm pots filled with pure silt. The pots were placed in the greenhouse at 20°C ± 3°C. After eight weeks, pots were inoculated with 20,000 J2 of *H. schachtii* (origin = Münster, Germany). After one life-cycle (approx. 6 weeks) was completed and female cysts had a brown color, the pots were harvested. Upper plant shoots were discarded and the contents of the pot were sifted through a sieve with a mesh of 7 mm. Then the soil containing cysts was mixed thoroughly. Before storing the cyst inoculum soil in plastic bags at 4°C, 50 g soil samples were taken to determine the average amount of cysts per 50 g of soil and the average amount of eggs and juveniles within the cysts.

To determine the amount of cysts per 50 g of soil and the number of eggs and juveniles within the cysts, they had to be extracted from the soil sample. With a water sprayer, the sample was passed through a 1 mm-sieve into a 5 l-bucket with a 250 µm aperture gauze at the bottom. The cysts retained on the gauze in the bucket were rinsed onto a 50 µm aperture sieve used for analysis and then rinsed into 250 ml glass beakers. A folded round paper filter (Schleicher & Schüll GmbH, Dassel, Germany Ø 185 mm) was placed in a glass funnel which was set in a 1 l-Erlenmeyer flask. Water containing cysts was passed through the filter, whereas cysts remained on the filter. Hereafter, the cysts were counted under a dissecting microscope, removed and filled into a 75 ml-plastic centrifuge-tube. Cysts were crushed with a metal cyst mill (RW 20n, IKA Company GmbH & Co. KG, Staufen, Germany) to release eggs and juveniles of *H. schachtii* (Goffart, 1958, Müller, 1980). The contents of the tube were rinsed into a 100 ml measuring cylinder filled with tap water up to 80 ml. One millilitre of the solution which had been pipetted into a 1 ml counting chamber was determined under a microscope and multiplied by 80 to obtain the number of eggs and juveniles per 50 g of soil. Fifty grams of cyst inoculum soil contained 93,760 eggs and larvae of *H. schachtii*.

2.2.2 Preparation of juvenile inoculum of Heterodera schachtii

To obtain juvenile inoculum of *H. schachtii* for trials, soil inoculum containing cysts of this nematode was required and prepared according to 2.2.1 Preparation of cyst inoculum of *Heterodera schachtii*. Depending on the number of juveniles needed, an estimated amount of soil inoculum was taken, passed through a 1 mm aperture sieve with a water sprayer and collected on a 250 µm aperture gauze at the bottom of a 5 l bucket. The cysts with a small amount of debris were rinsed with water from a fogger nozzle onto a 50 µm aperture sieve. Remaining residues were removed with water. Then cysts were divided among a sufficient number of milk filters which had been placed in 100 µm aperture sieves. (Hygia, Rapid, Paul Hartmann

AG, ø 140 mm) and set in Baermann funnels. The Baermann funnels were filled with a 0.04% ZnCl solution until the cysts were just covered. After three days at 24°C the juveniles (J2) were collected and their number was determined under a microscope (Leitz Labovert, Wetzlar, Germany). The suspension was then used as inoculum. Juveniles were stored for no longer than 1 week at 4°C and received fresh water every three days.

2.2.3 Preparation of juvenile inoculum of Meloidogyne incognita

Nematode inoculum was obtained from 2 to 3 month old tomato Lycopersicum esculentum ›Moneymaker‹ plants, which had been inoculated with 5000–20.000 J2 of M. incognita five weeks after seeding. The plants were kept in the greenhouse at 20°C ± 3°C and ambient lighting. Galled roots containing egg masses were washed free of soil with running tap water and cut into pieces 1 cm in length. The pieces of galled root were divided among a sufficient number of 100 μm mesh sieves, set in Baermann funnels and placed in a mist chamber (spray interval: 4.6 min., duration: approx. 34 sec.). Hatched juveniles were collected as of day 2. Juveniles were stored for no longer than 1 week at 4°C and received fresh water every three days.

2.3 Collection of experimental data

2.3.1 Root length/weight determination and enumeration of nematodes in a root system

The determination of root length or root weight respectively and the enumeration of nematodes in a root system were acquired in 100 ml biotests conducted with sugar beet against *H. schachtii* or tomato against *Meloidogyne* spp. Soil was removed from a seedling's roots by rinsing the roots under running tap water. Root length (cm) was recorded by measuring the length of the feeder root with a ruler. As this method does not account for the lateral roots, root weight was recorded with a scale (Sartorius 1205 MIO, K. Willers Laborbedarf, Münster, Germay) for later trials. Furthermore, the number of J2 of either *H. schachtii* or *M. incognita* in each root system was assessed. Therefore, the nematode infected plant roots were cleared by incubating them in 20% bleach (Danchlorix®, 5.25% sodium hypochlorite) for four minutes. The bleach was rinsed off the roots under running tap water and then soaked in tap water for 15 min. to remove any residual bleach which could have prevented the roots from staining. The roots were transferred into beakers with 40 ml of fresh tap water to which 1 ml of acid fuchsin stain solution was added. Stain solution was prepared by dissolving 3.5 g acid fuchsin in 250 ml acetic acid and 750 ml distilled water. The solution was put into a microwave oven for two minutes at the highest setting for it to boil. After the solution had cooled down to room temperature, the roots were rinsed and placed into dishes with glycerine. The dishes were placed in the refrigerator overnight for destaining (Hussey and Barker, 1973; Byrd et al. 1983). For examination the roots covered with glycerin were positioned between two microscopic glass slides and slightly pressed. The slides were then examined under a dissecting microscope for nematode numbers within in a root system.

2.3.2 Final population density determination of Heterodera schachtii

Evaluation of greenhouse pot trials with sugar beet and *H. schachtii* included the determination of the final population density (Pf) after a completed life-cycle of *H. schachtii* (Greco, 1982). Cysts were extracted from soil by means of centrifugation according to Caveness and Jenson (1955). For this, 200 g soil samples containing cysts of *H. schachtii* were taken from each pot, passed through a 1 mm aperture kitchen sieve with a water sprayer and collected on a 250 μm aperture gauze within a 5 l bucket. Cysts mixed with soil debris were rinsed into 1 l centrifuge beakers with water from a fogger nozzle. After adding two tablespoons of Kaolin to the mixture, the beakers were placed in a centrifuge (Cryofuge 6-6, Heraeus Christ, Hanau, Germany) for 5 min at 3000 rpm to separate water from the soil containing the cysts. Water with debris was poured onto a 50 μm aperture sieve. Hereafter, a solution of $MgSO_4$ (density = 1.2) was added to separate the cysts of *H. schachtii* from the soil. The contents of the beaker were mixed with a vibromixer (Vibromixer type E, Chemap AG, Männedorf, Switzerland) and placed in the centrifuge for another five minutes at 3000 rpm. $MgSO_4$ containing cysts of *H. schachtii* were poured onto the 50 μm aperture sieve and rinsed into 250 ml glass beakers with water from a fogger nozzle. A folded round paper filter (Schleicher & Schüll GmbH, Dassel, Germany Ø 185 mm) was placed in a glass funnel which was set in an Erlenmeyer flask. Water containing cysts was passed through the filter retaining cysts. Hereafter, the cysts were counted under a dissecting microscope, removed and filled into a 75 ml plastic centrifuge-tube. Cysts were crushed with a metal cyst mill (RW 20n, IKA Company GmbH & Co. KG, Staufen, Germany) to release eggs and juveniles of *H. schachtii* (Goffart, 1958, Müller, 1980). After the contents of the tube were filled into a 100 ml measuring cylinder the solution was diluted to 80 ml. The number of eggs and juveniles within 1 ml was quantified and multiplied by 80 to obtain the number of eggs and juveniles within a 50 g soil subsample.

2.3.3 Extraction of free-living second-stage juveniles

The extraction of free-living second-stage juveniles of *H. schachtii* and *Meloidogyne incognita* is similar to the extraction of cysts of *H. schachtii* described previously. Briefly, 200 g soil samples were placed in 1 l centrifuge beakers and covered with water. The beakers were then placed in the centrifuge for 5 min at 3000 rpm. Water with bits of debris was tossed and $MgSO_4$-solution with a density of 1.15 was added to the beakers to separate second-stage juveniles from soil particles. The contents of the beaker were mixed thoroughly with a vibromixer for one minute. Again, the beakers were placed in the centrifuge for 5 min. at 3000 rpm. To retrieve second-stage juveniles, $MgSO_4$-solution containing second-stage juveniles were passed through a 20 μm aperture sieve. The collected juveniles from one beaker were rinsed into a 100 ml measuring cylinder with water from a fogger nozzle. The juvenile suspension was diluted to 80 ml. One millilitre of suspension was filled into a 1 ml counting chamber and the number of juveniles within 1 ml were counted under a microscope and multiplied by 80 to obtain the number of juveniles within a sample.

2.3.4 Cultivation and conservation of Hirsutella rhossiliensis

Hirsutella rhossiliensis isolate BBA was originally isolated from second-stage juveniles of *Heterodera schachtii* in 1985 at the Federal Biological Research Centre for Agriculture and Forestry (BBA), Münster, Germany. The juveniles were derived from microplots, in which sugar beet was grown. The fungus was maintained on beads of a Cryobank system™ (Mast Diagnostica, Laboratoriumspräparate GmbH, Reinfeld, Germany) and stored at minus 80°C. To obtain sporulating aerial mycelia of the fungus, it was cultured on PDA at 23°C for two weeks to allow adequate sporulation and stored in the refrigerator for no longer than two months.

2.3.5 Fermentation and encapsulation of Hirsutella rhossiliensis

Biomass (vegetative hyphae) of *H. rhossiliensis* needed for encapsulation was produced by BIOCARE GmbH, Einbeck, Germany, whereas biomass needed for trials without a capsule treatment was produced at the BBA in Münster. Fungal biomass was prepared in an optimized and modified liquid media according to Rose (2000). Components of the liquid media are listed in the table below (Table 1).

Table 1: Modified liquid media for culture of Hirsutella rhossilienisis *according to Rose 2000.*

Components	% (w/w)
Glucose	2.0
Yeast extract	0.5
MES-NaOH-Buffer	0.4
K_2HPO_4	0.1
Penicillin	0.0025
Streptomycin	0.0025
Destilled water	96.595

The fungus was transferred from a culture on PDA which was provided by the BBA Münster (see chapter 2.3.4 Cultivation and conservation of *Hirsutella rhossiliensis* for details) to each of six 1 l Erlenmeyer flasks containing 500 ml of sterile liquid media. The flasks were shaken on an orbital shaker at 120 rpm for 12 days at room temperature (21°C-23°C). The fungal colony pellets were recovered by filtering the media containing mycelia of *H. rhossiliensis* over a black band filter 589[1] (Schwarzband, Dassel, Germany) with a membrane vacuum pump (KNF, Neuberger Laboport GmbH, Freiburg, Germany). The fungal colony pellets were rinsed with sterile distilled water three times to remove nutrient solution. To obtain sufficient biomass for capsules applied in larger greenhouse trials, biomass, harvested from liquid shake culture was multiplied in a bioreactor at the company BIOCARE GmbH, Einbeck, Germany (Patel et al., 2004). After biomass was harvested, 1%, 5% or 10% mycelium of *H. rhossiliensis* was added to encapsulation liquid containing hydrogel-forming biopolymers based on renewable resources and 3% technical yeast extract (baker's yeast), dropped with the Jet Cutter into a 2% $CaCl_2$-solution and left to crosslink for 20 min. Capsules were packed on ice and sent to the Federal Biological

Research Centre for Agriculture and Forestry (BBA), Germany and arrived within one to two days.

2.4 Statistical Analysis

Data were analysed according to standard analysis of variance procedures with the SPSS 11 program for Windows. Differences among treatments were tested using a one way analysis of variance (ANOVA) followed by Tukey Test HSD (Homogenous Significant Difference) for mean comparison if the F-value was significant and data had a normal distribution. Statistical differences referred to in the text were significant at a probability of error of 5% ($p \leq 0.05$). Interactions among treatments were tested using a two way analysis of variance followed by Holm-Sidak (All Pairwise Multiple Comparison Procedures). Statistical differences referred to in the text were significant at a probability of error of 1% ($p \leq 0.001$) and 5% ($p \leq 0.05$). However, T-test was used for comparing 2 treatments using Windows Excel.

3 Host suitability of the *Hirsutella rhossiliensis* isolate BBA

3.1 Introduction

In this study the endoparasite *H. rhossiliensis* was investigated for its host suitability towards plant-parasitic nematodes but also its side-effect on non-target organisms e. g. entomogenous nematodes. In the past, this fungus has been isolated from various nematodes and locations and has proven a good ability to parasitize a broad range of plant-parasitic nematodes such as representatives of the genera *Globodera, Heterodera, Meloidogyne, Pratylenchus* and *Ditylenchus* (Núñez-Fernández, 1992; Cayrol et al., 1986; Timper & Brodie, 1993). It is known that isolates of fungi can differ in their pathogenicity and virulence (Irving and Kerry, 1986; McCoy, 1990) and researchers have shown that isolates of *H. rhossiliensis* may vary considerably in virulence and pathogenicity to certain nematode species (Sturhan and Schneider, 1980; Jaffee and Zehr, 1985; Jaffee et al., 1991; Timper and Brodie, 1993; Velvis and Kamp, 1995; Liu and Chen, 2000). Variability among isolates of *H. rhossiliensis* has been demonstrated by Tedford et al., 1994. He showed that for instance isolates from nematodes in the *Hoplolaimidae* (*Rotylenchus robustus* and *Hoplolaimus galeatus*) grew slower on agar, produced larger conidia, and were less pathogenic towards nematodes than were isolates from other hosts. The objective of this study was to compare the pathogenicity of *H. rhossiliensis* isolate BBA to *H. schachtii* and other nematode species. The nematode species chosen for this study could be classified into three groups: sedentary endoparasites, migratory endoparasites and entomogenous nematodes. Sedentary parasites chosen for this trial were the root-knot nematodes *Meloidogyne incognita* and *M. hapla* as well as the sugar beet cyst nematode *H. schachtii*. Other economically important pests are migratory endoparasites belonging to the genus *Pratylenchus* and *Radopholus*. The species *P. penetrans*, a pest of a wide variety of crops and *R. similis*, an important pest of banana, were chosen for this study. Of great interest is not only the pathogenicity of *H. rhossiliensis* isolate BBA towards plant-parasitic nematodes but also its side-effect on non-target organisms such as the entomogenous nematodes *Steinernema feltiae* and *Heterorhabditis bacteriophora*, which are important antagonists of soil-dwelling insects (Kaya, 1985). For all selected nematode species, the pathogenicity of *H. rhossiliensis* and the relationship between exposure time to the fungus and nematode mortality were investigated.

3.2 Material and Methods

3.2.1 Hirsutella rhossilensis

H. rhossiliensis isolate BBA was cultured on potato dextrose agar (PDA) at 24°C. The strain sporulated abundantly after fourteen days and was ready for testing.

3.2.2 Sedentary endoparasites

Heterodera schachtii

Heterodera schachtii was obtained from cysts that were extracted from pot cultures with fodder radish ›Siletina‹ (*Raphanus sativus oleiformis*). The cysts were incubated on Baermann funnels. The hatched second-stage juveniles (J2's) were collected in a glass beaker and received fresh water every day. They were stored at 6°C until sufficient numbers were collected but for no longer than a week. Juveniles were surface sterilized by incubating them for one minute in a 0.02% mercury solution followed by three washes in sterile water. About 1000 J2's of *H. schachtii* in 100 µl of sterile water were then transferred onto each of six PDA plates with sporulating *H. rhossiliensis*. The juveniles remained on three of the six Petri plates for one hour and on the other three plates for twenty-four hours at 23°C, respectively. Then the juveniles were rinsed off with sterile water into empty Petri plates and examined under a Leica MZ 8 dissecting microscope (Leica, Munich, Germany) for fungal infection. The dissecting microscope was set on a clean bench (Prettl, Bempflingen, Germany) to avoid external contamination. Individuals with attached conidia were transferred to Petri plates containing WA with a sterile glass pipette. Nematodes could easily be observed on the agar and there position was marked daily on the bottom of the plate with a different colour marker to see exactly when nematode movement ended and were compared to control plates containing non-infected nematodes. In addition, juveniles were considered dead when they showed no reaction after stimulation with a needle.

Meloidogyne incognita and M. hapla

Meloidogyne incognita and *M. hapla* were cultured on tomato plants ›UC 82 Davis‹ (*Lycopersicon esculentum*). Tomato roots containing egg masses of *M. hapla* or *M. incognita* were incubated on Baermann funnels set in a mist chamber at 22°C ± 3°C. The hatched second-stage juveniles of *M. incognita* and *M. hapla* were collected in a glass beaker and received fresh water every day. They were stored at 6°C until a sufficient amount was collected but for no longer than a week. Trial set-up was as described previously for *H. schachtii*. A variation was made in the amount of time the second-stage juveniles were exposed to sporulating *H. rhossiliensis*. Juveniles were exposed to the fungus for either two or twenty-four hours at 23°C. Nematode mortality was recorded as described for *H. schachtii*.

3.2.3 Migratory endoparasites

Pratylenchus penetrans

Pratylenchus penetrans was cultured on maize ›Oldham‹ (*Zea mays*) (Syngenta Seeds GmbH, Kleve, Germany). Maize roots containing adults and juveniles of *P. penetrans* were cut into 1 cm long pieces and incubated on Baermann funnels, which were set in a mist chamber at 22°C ± 3°C. Juveniles and adults that abandoned the root system were collected in a glass beaker and received fresh water every day. They were stored at 6°C until a sufficient amount was collected but for no longer than a week. Trial set-up was as previously described for *Heterodera schachtii* with a few modifications. *H. rhossiliensis* was cultured on 7.6% instead of 3.8% PDA in Petri plates at 24°C. A higher percentage of PDA was chosen to hinder the migratory nematodes from penetrating into the agar. *H. rhossiliensis* sporulated abundantly after fourteen days and was ready for testing. About 3000 juveniles and adults of *P. penetrans* in 75 µl of sterile water were transferred onto each of eight Petri plates with sporulating *H. rhossiliensis*. The nematodes remained onto the Petri dishes for three or twenty-four hours at 23°C. Nematode mortality was recorded as described for *H. schachtii*.

Radopholus similis

Pure cultures of the migratory endoparasite *Radopholus similis* on carrot discs were obtained from Prof. R. A. Sikora, University of Bonn, Germany. Nematodes of all developmental stages were rinsed from the carrot discs and collected in a beaker. Nematode sterilization was as described for *H. schachtii* and evaluation followed the procedure as described for *P. penetrans*.

3.2.4 Entomopathogenic nematodes

Steinernema feltiae

Pure cultures of the entomogenous nematode *S. feltiae* formulated in clay minerals were obtained from E-Nema, Raisdorf, Germany. The product Nemaplus® contained dauer larvae (third-stage juveniles) of *S. feltiae*, which is the infective and environmentally resistant stage. The body of the dauer larvae is ensheathed in a second-stage cuticle (Poinar and Leutenegger, 1968). Dauer larvae were incubated on Baermann funnels. Juveniles of *S. feltiae* were collected in a glass beaker and received fresh water daily. They were stored at 6°C until a sufficient amount was collected but for no longer than a week. Trial set-up followed the procedure described for *Heterodera schachtii*. Three-thousand dauer larvae of *S. feltiae* in 75 µl of sterile water were transferred onto each of nine PDA dishes with sporulating *H. rhossiliensis*. The dauer larvae were incubated on each set of three Petri plates for 10 min., 1 h and 24 h, respectively. Nematode mortality was recorded as described for *H. schachtii*.

Heterorhabditis bacteriophora

The product Nematop® contained pure cultures of the entomogenous nematode *H. bacteriphora* which were formulated in clay minerals. It was also obtained from E-Nema, Raisdorf, Germany. The dauer larvae, also ensheathed in a second-stage cuticle were incubated on Baermann funnels. They were stored at 6°C until a sufficient amount was collected but for no longer than a week. Trial set-up was as described previously for *H. schachtii*. Approximately 3000 dauer larvae of *H. bacteriophora* in 60 µl of sterile water were transferred onto each of six PDA plates with sporulating

H. rhossiliensis. The dauer larvae, on half of the Petri plates, were exposed to *H. rhossiliensis* for one hour and on the other three plates for 24 hours at 23°C, respectively. Nematode mortality was recorded as described for *H. schachtii.*

3.3 Results

Adherence of conidia to the cuticle of all nematode species occurred predominantly around the head.

3.3.1 Sedentary endoparasites

Heterodera schachtii, Meloidogyne incognita, M. hapla

The adhesion of conidia on the cuticle of *H. schachtii,* could be observed after one and twenty-four hours of exposure to sporulating aerial mycelia of *H. rhossiliensis.* *H. schachtii* exposed to *H. rhossiliensis* for one hour contained 10 to 30 adhesive conidia whereas juveniles exposed for twenty-four hours contained 20 to 40 adhesive conidia (Figure 3). Ten infected second-stage juveniles of *H. schachtii,* exposed to *H. rhossiliensis* for one hour or 24 hours respectively, were transferred onto each of five WA plates. Over 85% of infected *H. schachtii* juveniles, exposed to *H. rhossiliensis* for one hour, were killed by the fungus within two days, whereas mortality of non-treated juveniles was zero. Mortality of *H. schachtii* juveniles exposed to *H. rhossiliensis* for 24 hours was above 80% after two days and reached 96% after four days (Figure 2). For *M. incognita* and *M. hapla*, no attachment of fungal conidia was observed after two hours of incubation. From juveniles exposed for twenty-four hours, 33 J2's of *M. incognita* with 10 to 20 adhering conidia were found and transferred to four WA plates and 14 J2's of *M. hapla* with 1 to 5 adhering conidia were transferred to two WA plates. Within two days nematode mortality of *M. incognita* was above 80%, after four days it reached 87%, whereas nematode mortality of *M. hapla* reached only 43% after four days (Figure 2, Figure 4). Mortality of non-treated juveniles of *H. schachtii, M. incognita* and *M. hapla* was not observed within this experiment.

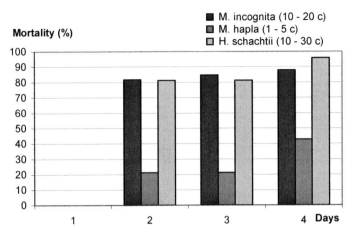

Figure 2: Relative mortality of Meloidogyne incognita *juveniles with 10 to 20 adhesive conidia (c)*, M. hapla *juveniles with 1 to 5 adhesive conidia (c) and* Heterodera schachtii *juveniles with 10 to 30 adhesive conidia (c) following exposure to sporulating aerial mycelia of* Hirsutella rhossiliensis *isolate BBA for 24 hours.*

Figure 3: *From left to right, second-stage juveniles of* Heterodera schachtii, Meloidogyne incognita *and* M. hapla *with adhesive conidia of* Hirsutella rhossiliensis *of isolate BBA after 24 h exposure (x 100).*

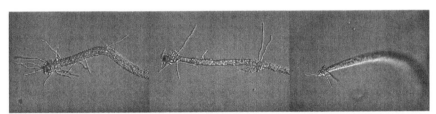

Figure 4: *From left to right, parasitized juveniles of* Heterodera schachtii, Meloidogyne incognita *and* M. hapla *four days after 24 hour exposure to* Hirsutella rhossiliensis *isolate BBA (x 25).*

3.3.2 Migratory endoparasites

Pratylenchus penetrans

From nematodes exposed for three hours to *H. rhossiliensis*, 21 juveniles and adults of *P. penetrans* with 2 to 20 adhering conidia were found and transferred onto WA. From nematodes exposed for twenty-four hours, 17 nematodes with 5 to 30 adhering conidia were found and transferred onto three WA plates. Five days after exposure to *H. rhossiliensis* the mortality rate of *P. penetrans* surpassed 75% regardless if nematodes had been exposed to the fungus for three hours or 24 hours (Figure 5). The remaining 25% were not infected by the adhering conidia. Natural mortality of *P. penetrans* was not observed after five days under the conditions of this experiment.

Figure 5: Mortality (%) of Pratylenchus penetrans *exposed to sporulating aerial mycelia of* Hirsutella rhossiliensis *isolate BBA for 3 hours resulting in 2 to 20 adhesive conidia (c) and 24 hours resulting in 5 to 30 adhesive conidia (c).*

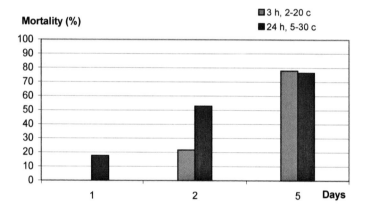

Radopholus similis

No adhesion of conidia to the cuticle of juveniles and adults of *R. similis* was observed within three hours of exposure to *H. rhossiliensis*. After 24 hours of exposure to *H. rhossiliensis* a group of 25 nematodes with 5 to 15 conidia as well as a group of 28 nematodes with 15 to 30 adhering conidia were found and transferred onto WA. *R. similis* with 5 to 15 conidia exhibited a mortality of 24% after two days and 76% after five days. A higher amount of conidia led to an increase of mortality. *R. similis* with 15 to 30 adhering conidia showed a mortality of 46% after two days and 86% after five days (Figure 6). Natural mortality of *R. similis* was not observed after five days under the conditions of this experiment.

Mortality (%)

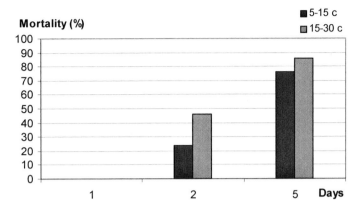

■ 5-15 c
▨ 15-30 c

Figure 6: Mortality (%) of Radopholus similis *with 5 to 15 adhesive conidia (c) or 15 to 30 adhesive conidia after exposure to sporulating aerial mycelia of* Hirsutella rhossiliensis *isolate BBA for 24 hours.*

3.3.3 Entomopathogenic nematodes

Steinernema feltiae

Dauer larvae of Steinernema feltiae exposed to sporulating H. rhossiliensis isolate BBA for ten minutes or one hour had no adhering conidia. From nematodes exposed for twenty-four hours, 60 third-stage juveniles with 5 to 10 adhering conidia were found and transferred onto six WA plates and 40 juveniles with 10 to 20 adhering conidia to four WA plates (Figure 7). Within four days on WA more than 96% of *S. feltiae* were infected and killed regardless of the number of adhesive conidia (Figure 8).

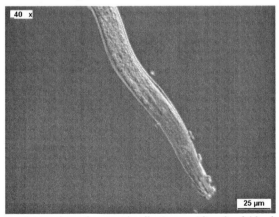

Figure 7: Steinernema feltiae *dauer larva with adhering conidia after 24 h of exposure to sporulating aerial mycelia of* Hirsutella rhossiliensis *isolate BBA.*

39

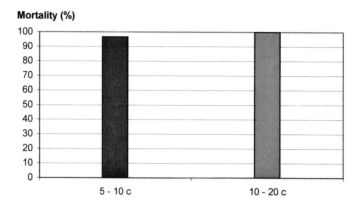

Figure 8: Mortality (%) of Steinernema feltiae *dauer larvae, exposed to sporulating aerial myce-lia of Hirsutella rhossiliensis isolate BBA for 24 hours resulting in 5 to 10 adhesive conidia (c) or 10 to 20 adhesive conidia (c), after four days.*

Natural mortality of *S. feltiae* was not observed after four days under the conditions of this experiment.

Heterorhabditis bacteriophora

After one hour of exposure to sporulating aerial mycelia of *H. rhossiliensis*, 50 *H. bacteriophora* dauer larvae with 1 to 10 adhesive conidia of *H. rhossiliensis* were transferred onto 5 WA plates. Although nematodes had attached conidia, only 1% of the dauer larvae were parasitized and killed after nine days. Cast J2 cuticles with attached conidia could be observed on the Petri plates filled with 1.5% WA. After exposure to the fungus for 24 hours, 51 third-stage juveniles with 10 to 20 adhering conidia were transferred to five WA plates (Figure 9). The mortality rate of *H. bacteriophora* was below 20% after two days. Within six days about 55% of the third-stage juveniles were killed and after nine days the mortality rate was 68% (Figure 10).

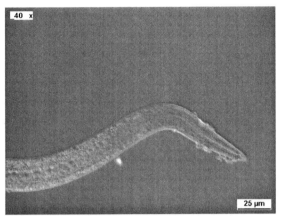

Figure 9: Heterorhabditis bacteriophora *dauer larva with adhesive conidia after 24 h of exposure to sporulating aerial mycelia of* Hirsutella rhossiliensis *isolate BBA.*

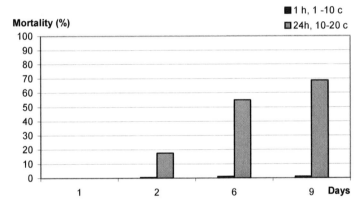

Figure 10: Mortality (%) of Heterorhabditis bacteriophora *dauer larvae exposed to* Hirsutella rhossiliensis *for 1 hour with 1 to 10 adhesive conidia (c) and 24 hours with 10 to 20 adhesive conidia (c) 1, 2, 6 and 9 days after exposure to sporulating aerial mycelia of* H. rhossiliensis *isolate BBA.*

3.4 Discussion

This study demonstrated that *H. rhossiliensis* isolate BBA has a broad host spectrum although highest mortality was achieved on its original host *H. schachtii*. After one hour of exposure to *H. rhossiliensis* isolate BBA, *H. schachtii* juveniles contained up to 30 adhesive conidia and after transfer to WA, 85% of the juveniles were killed within two days. No other species tested showed such a high amount of adhering conidia within such a short time. The sedentary endoparasite *M. hapla* and *M. incognita* did not have adhesive conidia after one hour of exposure to the fungus, although they are known to be good hosts for *H. rhossiliensis* (Cayrol et al., 1986; Tedford et al., 1992). An explanation could be limited nematode movement of

Meloidogyne spp. compared with *H. schachtii*. When observed in water *H. schachtii* was more active than *Meloidogyne* spp. As suggested by Timper et al. (1991), nematodes must move to encounter *H. rhossiliensis* conidia, therefore differences in mobility among nematode species consequently would result in differences in the number of conidia encountered. Thus extending the exposure time to the fungus for 24 hours would increase the infection rate which was observed in our study. After 24 hours of incubation, second-stage juveniles of *M. hapla* with only 1 to 5 adhesive conidia and juveniles of *M. incognita* with 10 to 20 adhesive conidia were observed. An explanation for the range of infection rates among the different species could be cuticular differences among the various species. Timper et al. (1991) suggested that adhesion of conidia to nematode cuticles is not an »all or none response,« but that there may be a range of adhesiveness depending upon the species of the nematode and the isolate of the fungus. Of the migratory endoparasites *R. similis* and *P. penetrans*, incubated on the Petri dishes with sporulating aerial mycelia of *H. rhossiliensis* for three hours, only a fraction of 21 nematodes of *P. penetrans* acquired conidia. Of the entomogenous nematodes *S. feltiae* and *H. bacteriophora*, only *H. bacteriophora* acquired conidia for which a group of 50 nematodes with adhesive conidia were found after one hour of exposure to the fungus. Timper and Brodie (1993) observed that *H. bacteriophora* was more active than *P. penetrans* in water and that the low percentage of *P. penetrans* acquiring *H. rhossiliensis* conidia compared with *H. bacteriophora* was probably partly due to the lower mobility of *P. penetrans*. Although *H. bacteriophora* acquired conidia after three hours of incubation, only 1% of the dauer larvae were infected and killed by *H. rhossiliensis*. As the dauer larva is ensheathed in the second-stage cuticle (Poinar and Leutenegger, 1968), it is protected from the environment and from infection by *H. rhossiliensis*. Timper and Kaya (1989) reported that conidia of *H. rhossiliensis* adhered to the cuticle of *Heterorhabditis* spp. but less than 0.7% of the dauers became infected. Although conidia of *H. rhossiliensis* adhered to the J2 cuticles of *H. bacteriophora*, in most cases, they did not germinate. Occasionally, cast J2 cuticles with attached conidia were sometimes observed on WA in our study. Poinar and Jansson (1986) also observed that dauer larvae trapped by *Monacrosporium ellipsosporum* sometimes escaped infection by slipping out of their J2 cuticle. Perhaps this was an additional method to escape infection. In our study conidia of *H. rhossiliensis* did not adhere to the J2 cuticle of *S. feltiae* after one hour of exposure to the fungus. This observation was also made by Timper and Kaya (1989), although conidia did adhere to the J3 cuticle if dauer larvae had lost their J2 cuticles. If *H. bacteriophora* and *S. feltiae* remained on PDA with *H. rhossiliensis* for 24 hours, often their J2 cuticles became exsheathed and the species became more susceptible for conidia adhesion and infection. Timper and Kaya (1989) demonstrated that *Steinernema* spp. lost their cuticles after moving through 5 cm of moist sand, whereas *Heterorhabditis* spp. generally retained their J2 cuticles. In our laboratory assay, *H. bacteriophora* perhaps exsheathed its J2 cuticle as a protective measure and was infested anew, whereas *S. feltiae* readily lost its cuticle within the 24 hours of incubation on PDA with *H. rhossiliensis*. As it has been reported that 12 hours are required for 50% of the conidia to penetrate and form infection bulbs in *H. schachtii* at optimal temperature of 25°C (Tedford et al., 1995), it is assumed that *H. bacteriophora* and *S. feltiae* could not avoid adherence and infection as they remained on the dishes for 24 hours. Furthermore, if incubated for 24 hours, the number of adhering conidia in the head region increased strongly which could have led to an infection of *H. bacteriophora* larvae with retaining J2. Timper and Kaya (1989) also observed that infections occurred around the head, although this occurred rarely.

Transmission of spores and infection of all nematode species were successful if nematodes remained on the Petri dishes with sporulating aerial mycelia of *H. rhossiliensis* for at least 24 hours. Approximately 70% of *H. bacteriophora* were killed within 9 days after exposure to the fungus for 24 hours. A mortality rate of about 80% was reached by the species *M. incognita, P. penetrans* and *R. similis* after four to five days and a mortality rate of over 90% was reached by *Steinernema feltiae* after four days. Under field conditions nematodes are exposed to the fungus for only a short period of time as they are not trapped on the fungus but continue to move throughout the soil, which makes infection harder. But movement of plant-parasitic nematodes might increase in the presence of attractive host plants, thus promoting transmission of *H. rhossiliensis* to the nematodes. The entomopathogenic nematode species *S. feltiae* can escape infection by *H. rhossiliensis* in the field by slipping out of its protective J2 cuticle with adhesive conidia of *H. rhossiliensis*, whereas conidia of *H. rhossiliensis* that adhere to the J2 cuticle of *H. bacteriophora* normally do not germinate. If, in fact, an infection occurs after the dauer larva has exsheathed its protective J2 cuticle, parasitism can take place. Generally though, most dauer larvae retain their J2 cuticles and are therefore protected against infection.

3.5 Summary

Hirsutella rhossiliensis isolate BBA was tested for its pathogenicity to *Heterodera schachtii, Meloidogyne incognita, M. hapla, Pratylenchus penetrans, Radopholus similis* and its side-effects on non-target organisms *Steinernema feltiae* and *Heterorhabditis bacteriophora* in laboratory assays. The different species were exposed to the fungus between ten minutes and twenty-four hours. After one to three hours of exposure to the fungus the plant-parasitic nematode species *M. incognita, M. hapla, R. similis* did not reveal adhesive conidia. Possible explanations are lower mobility of the nematode species or low susceptibility. Extended exposure for 24 hours to the fungus proceeded in adhesion of conidia to the cuticle and infection and nematode mortality of over seventy percent among all plant-parasitic nematode species. *H. schachtii* showed best results concerning conidia adherence and nematode mortality. The non-target organism *S. feltiae* did not show adhesion of conidia to the cuticle after one hour of exposure to the fungus, whereas *H. bacteriophora* did. Although juveniles with adhering conidia were found, only 1% of *H. bacteriophora* dauer larvae were infected and killed by *H. rhossiliesis*. The retaining J2 cuticle of these entomopathogenic nematode species protects the nematode from infection.

4 Influence of plant protection agents on the growth of *Hirsutella rhossiliensis*

4.1 Introduction

The biocontrol effectiveness of *Hirsutella rhossiliensis* against the two important plant-parasitic nematode species *Heterodera schachtii* and *Meloidogyne incognita* was investigated within this research. The sugar beet cyst nematode *H. schachtii* has remained one of sugar beet's (*Beta vulgaris*) most damaging pests and occurs in all major sugar beet growing areas of Central Europe (Cooke, 1993). In temperate regions, *M. incognita* predominantly occurs in greenhouses where it can cause economic damage especially on tomato, cucumber and pepper. Narrow rotations greatly favour nematode increase (Richardson and Grewal, 1993). In Germany, all but one nematicide have been removed from the market due to environmental concerns. The remaining nematicide, Nemathorin®, is only registered for use in late potatoes. For other crops, nematode control strategies completely depend on an integrated approach. A possible approach could be the incorporation of the nematophagous fungus *H. rhossiliensis* as a biocontrol agent in field or greenhouse soil. Unfortunately, besides nematodes, crops are attacked by several other biotic factors, such as pathogens, weeds and insects. Those biotic factors are generally controlled by single or repeated treatments with plant protection agents that are applied throughout the vegetation season. Until now, little is known about the side-effects of plant protection agents on biocontrol agents. The usage of *H. rhossiliensis* as a biological control agent can only be of advantage to the grower if this fungus is compatible with plant protection agents used to secure plant health and yield. The objective of this study was to evaluate the compatibility of *H. rhossiliensis* to five fungicides, five insecticides and four herbicides commonly used in sugar beet production and greenhouse production of tomato and cucumber. We hypothesized that insecticides and herbicides would not have side-effects on our fungus as their mode of action is much different to that of fungicides but that perhaps fungicides might have affect the growth of *H. rhossiliensis*.

4.2 Material and Methods

4.2.1 Experimental Design

An *in vitro* test was conducted to determine the effects of selected plant protection agents on growth of colony forming units of *H. rhossiliensis* from a fungal suspension. The fungal suspension consisted of 0.2% (w/w) vegetative hyphae *of H. rhossiliensis*. Vegetative hyphae of *H. rhossiliensis* were collected as described in 2.3.5 »Fermentation and encapsulation of *Hirsutella rhossiliensis*«. A dilution series was created for each pesticidal solution tested which also contained 10% fungal suspension. The pesticidal solution was tested at the recommended application rate (100%), a 10-fold (10%) and 100-fold (1%) dilution. A control (0%) was added to each pesticide tested. One-hundred µl of each suspension of the dilution series was applied onto PDA plates and spread evenly. Each treatment had five replicates. Plates were incubated at 24°C and the number of colony forming units was assessed after five days. The active ingredients of each product are listed in the tables below as well as the recommended application rate and the calculation based on 100 ml of sterile, distilled water (Table 2, Table 3, Table 4).

Table 2: Fungicides

Tradename, Manufacturer	Active ingredient, Target organism	Recommended application rate	Concentration in test (per 100 ml H_2O)
Aliette® (1)	Fosetyl (*Pseudoperonospora cubensis* on Cucumber)	4.5 kg/ha (4.5 kg diluted in 400 l of H_2O)	1125 mg
Euparen M WG® (1)	Tolylfluanid (*Phytophtora infestans* on tomato; *Sphaerotheca fuliginea* on cucumber)	5 g/Liter	500 mg
Bayfidan® (1)	Triadimenol (*Erysiphe betae* on sugar beet)	500 ml/ha (500 ml diluted in 400 l of H_2O)	0.125 ml
Previcur® (1)	Propacarb (*Pythium* on cucumber *Phytophtora* spp. on tomato in the greenhouse)	0.15%	0.15 ml
Cupravit ›Kupferkalk‹® (1)	Copper oxychloride (various fungal diseases on tomato)	3.15 kg/ha (3.15 kg diluted in 400 l of H_2O)	787.5 mg

(1): Bayer CropScience AG, Leverkusen, Germany

Table 3: Herbicides

Tradename, Manufacturer	Active ingredient, Target organism	Recommended application rate	Concentration in test (per 100 ml H_2O)
Goltix® (2)	Metamitron (selective herbicide used on sugar beet)	2 kg/ha (2 kg diluted in 400 l of H_2O)	500 mg
Betanal® Progress (1)	Phenmedipham, Desmedipham Ethofumesate, (selective herbicide used on sugar beet)	1.5 l/ha (1.5 l diluted in 400 l of H_2O)	0.375 ml
Basta® (1)	Glufosinat (non-selective herbicide used on sugar beet)	5 l/ha (5l diluted in 400 l of H_2O)	1.25 ml
Roundup® Ultra (3)	Glyfosat, non-selective herbicide used on cucumber in the field	4 l/ha (4 l diluted in 400 l of H_2O)	1 ml

(1): Bayer CropScience AG, Leverkusen, Germany
(2): Mauk Company, Thatcham, UK
(3): Monsanto Company, Düsseldorf, Germany

Table 4: Insecticides

Tradename, Manufacturer	Active ingredient, Target organism	Recommended application rate	Concentration in test (per 100 ml H_2O)
Bulldock® (1)	Beta-Cyfluthrin against aphids on sugar beet;	300 ml/ha (300 ml diluted in 400 l of H_2O)	0.075 ml
NeemAzal-T/S® (2)	Azadirachtin against *Tetranychus urticae* (spider mites) on decorative plants	3 l/ha (3 l diluted in 400 l of H_2O)	0.75 ml
Vertimec Abamectin® (3)	Abamectin against *Tetranychus urticae* (spider mites) on cucumber and tomato	0.9 l/ha (900 ml diluted in 900 l of H_2O)	0.1 ml
Metasystox R® (1)	Oxydemeton-methyl against aphids on sugar beet	600 ml/ha (600 ml diluted in 400 l of H_2O)	0.15 ml
Pirimor G WG® (1)	Pirimicarb against aphids on tomato and cucumber	0.5 g/l	50 mg

(1): Bayer CropScience AG, Leverkusen, Germany
(2): Trifolio-M Company, Lahnau, Germany
(3): Syngenta Company, Basil, Switzerland

4.3 Results

4.3.1 Fungicides

The products Previcur®, Aliette AG®, Bayfidan® und Cupravit® had no influence on the number of fungal colony forming units in comparison to their respective controls. Only Euparen M WG® fully inhibited the growth of colonies of the fungus *H. rhossiliensis* at the recommended application rate (100%) and at a dilution of 1:10 (10%). At a dilution of 1:100 (1%) the number of cfu/ml was reduced by 96% in comparison to the non-treated control (0%) for Euparen M WG® (Figure 11).

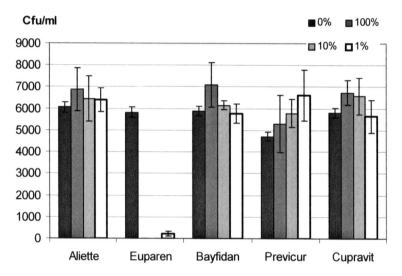

Figure 11: Influence of the fungicides Aliette®, Euparen®, Bayfidan®, Previcur® and Cupravit®, applied at the recommended application rate = 100%, diluted 1:10 = 10% and 1:100 = 1% on the number of colony forming units of the fungus Hirsutella rhossiliensis *per ml of suspension in comparison to the non-treated control (0%), means, bars indicate standard deviation, n = 5.*

4.3.2 Herbicides

The herbicides Roundup Ultra® and Goltix WG® had no influence on the growth of colony forming units of *H. rhossiliensis* (Figure 12). A treatment with Betanal Progress® and Basta® led to a reduction of colonies at the recommended application rate (100%) and at a dilution of 10%. A treatment of Betanal Progress®, applied at the recommended application rate, reduced the amount of colony forming units by more than 98% in comparison to the control (0%), whereas a dilution of 10% led to a reduction of colonies by 55% in comparison to the control. If applied at a dilution of 1%, the herbicide had no side-effect on *H. rhossiliensis*. A treatment of Basta®, applied at the recommended application rate, reduced the amount of colony forming units by more than 95% in comparison to the control. An application of this herbicide diluted to 10% reduced the number of colony forming units by more than 48%. A dilution of 1% had no effect on the number of colonies formed by *H. rhossiliensis* (Figure 12).

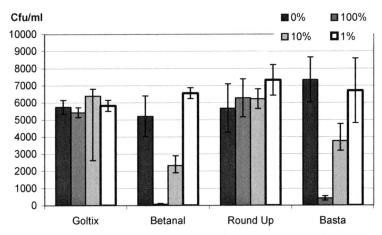

Figure 12: Influence of the herbicides Goltix WG®, Betanal Progress®, Roundup Ultra®, and Basta® applied at the recommended application rate = 100%, diluted 1:10 = 10% and 1:100 = 1% on the number of colony forming units of the fungus Hirsutella rhossiliensis *per ml of suspension in comparison to the non-treated control (0%), means, bars indicate standard deviation, n = 5.*

4.3.3 Insecticides

Three of five insecticides were incompatible with *H. rhossiliensis* at the recommended application rate. Although the insecticides Bulldock®, NeemAzal-T/S® and Metasystox® inhibited the growth of colonies of the fungus *H. rhossiliensis* at the suggested application rate, the inhibition did not reoccur at a dilution of 10% and 1%. Vertimec Abamectin® und Pirimor® did not have an effect on the number of fungal colony forming units in comparison to the non-treated controls of both insecticides, regardless of the concentration (Figure 13).

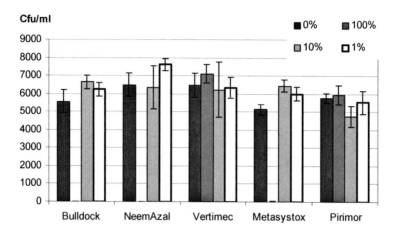

Figure 13: Influence of the insecticides Bulldock®, NeemAzal T/S®, Vertimec Abamectin®, Meta-systox® and Pirimor® applied at the recommended application rate = 100%, diluted 1:10 = 10% and 1:100 = 1% on the number of colony forming units of the fungus Hirsutella rhossiliensis *per ml of suspension in comparison to the non-treated control (0%), means, bars indicate standard deviation, n = 5.*

4.4 Discussion

As most chemical nematicides have been removed from the market in Germany, alternative methods for nematode control have to be developed. This research focused primarily on control of the sugar beet cyst nematode *H. schachtii* and secondarily on control of the root-knot nematode *Meloidogyne incognita* with the endoparasitic fungus *H. rhossiliensis*. The fungus was selected as it has been reported to suppress plant-parasitic nematodes and may face fewer restrictions towards widespread use and commercialisation. As the fungus has been isolated worldwide, it has adapted to survive in diverse environments such as field soil and glasshouse environments (Jaffee and Muldoon, 1989; Jaffee et al., 1989; Sturhan and Schneider, 1980). In this study, common plant protection agents, which are widely used in sugar beet production or in greenhouse production of tomato and cucumber, were chosen for evaluation. Recommended application rates plus a dilution of 1:10 (10%) and 1:100 (1%) were used to study their direct effect on growth of *H. rhossiliensis*. The dilutions were chosen to simulate potential soil water concentrations after application. The chosen insecticides and fungicides are applied to the foliar, so they usually do not come into direct contact with *H. rhossiliensis* in the soil. Similar, herbicides are applied to the soil and will only penetrate the soil in a diluted form. Therefore, only capsules lying on the surface are exposed to the full dosage of those plant protection agents. Of five fungicides tested only the fungicide Euparen M WG® suppressed the formation of *H. rhossiliensis* colony forming units *in vitro*. Euparen M WG®, containing the active ingredient Tolyfluanid, is categorized as an inhibiting, systemic fungicide that shows great activity against fungal metabolism and subsequent growth. On the other hand Bayfidan® and Cupravit ›Kupferkalk‹® are protectant fungicides that primarily inhibit spore germination and protect plants from fungal infection. This might perhaps explain why they did not affect *H. rhossiliensis*, which primarily occurs in soil in form of assimilative hyphae. Perhaps the growth of assimilative, vegetative hyphae of *H. rhossiliensis* was not disturbed by protectant fungicides. Of the few other studies investigating the side-effects of plant protection agents against nematophagous fungi, Woodward et al. (2005) conducted an *in vitro* tolerance assay with conidia of the nematophagous fungus *Arthrobotrys oligospora* and two fungicides commonly used against fungal diseases on putting greens. Chlorothalonil, a protectant fungicide and Myclobutanil, a sterol biosynthesis inhibiting systemic fungicide had adverse effects on *A. oligospora* spore germination and hyphal growth *in vitro*. Perhaps protective contact fungicides due to their mode of action do not effect hyphal growth but could effect sporulation of spores of *H. rhossiliensis* that have attached to plant-parasitic nematodes. It seems that some systemic curative fungicides such as Euparen M WG® have a greater impact on hyphal growth of *H. rhossiliensis* than protective contact fungicides. In this *in vitro* study the influence of plant protection agents on the germination of spores attached to the cuticles of plant-parasitic nematodes was not investigated. Jaffee and McInnis (1990) demonstrated in an *in vitro* test that the nematicide carbendazim (methyl benzimidazol-2-ylcarbamate; DPX 965-50DF; 50% a. i.; E. I. DuPont DeNemours and Co., Wilmington, DE, USA) suppressed the germination of *H. rhossiliensis* spores, attached to the cuticle of *Criconemella xenoplax*, thus hindering infection. The aspect that plant protection agents could suppress the sporulation of spores attached to nematodes should be further investigated.

Results of the laboratory test conducted with four different herbicides showed that two of four herbicides were incompatible with *H. rhossiliensis*. A treatment with Betanal Progress® and Basta® led to a reduction of colonies at the suggested application rate (100%) and at a 10% dilution. It was hypothesized that herbicides would not have side-effects on *H. rhossiliensis* as it is a fungal pathogen, not a weed, assuming that the mode of action of herbicides would not have side-effects on the growth of the fungus. To explain this phenomenon, detailed research on the inhibitory mode of action of herbicides on *H. rhossiliensis* would have to be conducted. Literature indicating similar results could not be found. As herbicides are sprayed directly on the surface of the soil, these results indicate that herbicides of all types of plant protection agents are undoubtedly the greatest cause for concern when it comes to integrated control with *H. rhossiliensis* and products should be tested in advance for their side-effects on the fungus.

The *in vitro* test with five different insecticides showed that although the products Bulldock®, NeemAzal-T/S® and Metasystox® inhibited the growth of colonies of the fungus *H. rhossiliensis* at the recommended application rate, the inhibition did not reoccur at 10% or 1%. The reason for an incompatibility with these three insecticides at the regular dosage cannot be explained. Such high dosages might have had a toxic effect on the fungus, subsequently inhibiting its growth. In praxis, insecticides are usually applied to foliage so the solution would not come into direct contact with *H.*

rhossiliensis, incorporated in the soil. Furthermore, pesticidal solutions get absorbed by soil particles if they are rinsed off the foliage by rain water or irrigation which makes it even more unlikely for the pesticidal solution to come into direct contact with the fungus. Initially, we had hypothesized that fungicides, of the three types of plant protection agents tested, would have the greatest impact on fungal growth. Contrary results were obtained by this experiment but the tests were only conducted *in vitro*. Woodward et al. (2005) demonstrated in a pot trial that despite the *in vitro* findings, the mycelium of *A. oligospora* may not be adversely affected in pot trials by the fungicides Chlorothalonil and Myclobutanil, as nematode populations were lowest in pots treated with the fungus and the fungicide and were highest in pots only treated with the fungicide. Bruckner (2002) found contradictory results between *in vitro* and *in situ* tests conducted with Contans WG for control of Sclerotinia diseases and numerous synthetic pesticides. Results obtained by these researchers indicate that *in vitro* findings need to be confirmed in pot trials. Moreover, pot trials with non-treated field soil have to be conducted to simulate the real conditions of the usage of *H. rhossiliensis* and its potential combination partners.

4.5 Summary

Plant-parasitic nematodes are damaging pests on numerous crops including sugar beet and greenhouse grown cucumber and tomato. Chemical approaches for nematode control are facing more and more restrictions while there is a need for alternative measures such as biological control agents. However, biological control agents should be compatible with integrated pest management systems including applications of synthetic plant protection agents. The endoparasitic fungus *Hirsutella rhossiliensis* is a potential organism for a biological control agent of widespread, economically important plant-parasitic nematode species. To secure the growth of this fungus, which is incorporated in soil, in an integrated pest management system, side-effects of synthetic plant protection agents on this fungus were investigated. Laboratory assays were conducted to determine the *in vitro* tolerance of the nematophagous fungus *Hirsutella rhossiliensis* to various fungicides, insecticides and herbicides applied on sugar beet and greenhouse cultures. Four of five fungicides proved to be compatible with *H. rhossiliensis*, whereas two of the four herbicides tested, reduced the amount of colony forming units of *H. rhossiliensis* at the suggested application rate in comparison to the non-treated control. Only two of five insecticides did not have side-effects on *H. rhossiliensis*. Results of this study indicate that applications of plant control agents used to manage fungal diseases, weeds and insect pests do not necessarily affect *H. rhossiliensis* adversely thus making *H. rhossiliensis* an attractive microbial organism for a biological control agent applied in integrated control.

5 Efficacy of biopolymers MG and MF for encapsulation of *Hirsutella rhossiliensis* to control the sugar beet cyst nematode *Heterodera schachtii*

5.1 Introduction

In Germany, no chemical nematicides are registered for the control of the sugar beet cyst nematode *Heterodera schachtii*. Measures such as wide crop rotations, the cultivation of resistant varieties and resistant green manure crops can lead to a reduction of the nematode population density. However, growers do not often practice wide crop rotations in view of economic aspects. Sowing of resistant green manure crops, such as certain cultivers of mustard and fodder radish, must be done before the middle of August in order to achieve nematode control which is not always possible after late harvest of the previous crops. When cultivating resistant sugar beet varieties, it is important to consider that they can lead to a selection of resistance breaking pathotypes of *H. schachtii*. Investigations of Müller (1992) have shown that virulent nematodes occur naturally in low numbers in the field and that these may reproduce strongly, if resistant sugar beet cultivars are cultivated. This can lead to damage of resistant varieties from the third sugar beet crop on. Another possible method for control of *H. schachtii* is the application of natural antagonists. Antagonists that occur naturally in field soil play an important role in the control of plant-parasitic nematodes. Biological control agents based on antagonistic fungi are applied in various countries (e. g. Bioact® against *Meloidogyne* spp. on tomato in the USA). Up until now a biological control agent against the sugar beet cyst nematode has not been developed. A possible candidate for such a control strategy could be *H. rhossiliensis*. For example, natural occurrence of the endoparasitic fungus *H. rhossiliensis* can reduce *H. schachtii* infestation of cabbage up to 77% (Jaffee and Muldoon, 1989). A prerequisite for an efficient biocontrol agent is an adequate formulation that enables a good establishment of *H. rhossiliensis* in the field and guarantees a safe application. As spores that have detached from fungal mycelia are no longer infective and do not remain vital for a long period of time, fungal mycelia was formulated in this project. Baker's yeast was added to promote fungal growth and increase its competitiveness towards soil microorganisms. Furthermore, the formulation itself should consist of biopolymers based on renewable resources. So far, renewable resources are applied as biodegradable package material or as a substrate for biotechnological processes. In integrated pest control, however, renewable resources have hardly been used despite numerous possibilities. The goal of this study was to test the efficacy of moist and dry formulations of *H. rhossiliensis* based on renewable resources against *H. schachtii* on sugar beet.

5.2 Material and Methods

5.2.1 Comparison of Guargum MG, Guargum MF and alginate capsule material

The biopolymers, chosen as capsule material for the formulation of the fungus *H. rhossiliensis*, were: Guargum MG, Guargum MF and alginate. The latter was chosen as comparable capsule material, as studies have already been conducted with this material (Lackey et al., 1993; Patel, 1998).

Treatments of the biotests were:

(1) Non-treated control (CON)
(2) Autoclaved baker's yeast (BY)
(3) Fungal biomass of *H. rhossiliensis* (H.r.)
(4) Alginate capsules (ALG)
(5) Guargum MG capsules (MG)
(6) Guargum MF capsules (MF)
(7) Alginate capsules with baker's yeast (ALG+BY)
(8) Guargum MG capsules with baker's yeast (MG+BY)
(9) Guargum MF capsules with baker's yeast (MF+BY)
(10) Alginate capsules with baker's yeast and *H. rhossiliensis* (ALG+BY+H.r.)
(11) Guargum MG capsules with baker's yeast and *H. rhossiliensis* (MG+BY+H.r.)
(12) Guargum MF capsules with baker's Yeast and *H. rhossiliensis* (MF+BY+H.r.)

A vitality test was conducted to secure product quality of capsules at trial begin. For this, thirty capsules containing *H. rhossiliensis* were placed on moist, autoclaved filter paper, on PDA and on WA, respectively. Purity and growth rate of *H. rhossiliensis* was evaluated.

For moist capsules, the biocontrol efficacy was investigated in two biotests against *H. schachtii* on sugar beet ›Penta‹ using a heat-treated and a non-treated field soil/sand mixture (2:1, w/w, pH=6). The trial was carried out according to Gutberlet (2000). Each treatment consisted of ten replicates. Transparent plastic containers made of polyvinyl chloride (40 x 20 x 120 mm, Kelder Plastibox b. v., s'-Heerenberg, The Netherlands) were filled with 100 ml of substrate. Capsules with and without fungus were thoroughly mixed into the soil (4 g/100 ml soil) at day 0. The treatment »Fungal biomass of *H. rhossiliensis*« consisted of fungal mycelium suspended in 4 ml of water per 100 ml of soil which was equivalent to the fungal amount in the capsules (1%). The treatment »Autoclaved baker's yeast« consisted of baker's yeast suspended in 4 ml of water and added to 100 ml of soil which was also equivalent to the yeast amount in the capsules (3%). The containers were set in the greenhouse at an average temperature of 20°C ± 3°C. Plants were watered as needed. At day 7, a sugar beet seed was sown 0.5 cm deep into the soil of each container. At day 14, sugar beet seedlings were inoculated with 1000 infective second-stage juveniles of *H. schachtii* in 2 ml of tap water. The juvenile suspension was pipetted into two one-centimetre-deep holes to the left and right of the sugar beet seedling. After an additional week, plants were extracted from the containers and the trial was evaluated. Plant shoots were discarded and roots rinsed free of soil debris under running tap water. The feeder root length was measured in cm and the juveniles within the root system were stained with a 0.01% acid fuchsin solution (Byrd et al., 1983) and quantified under a dissecting microscope.

5.3.2 Effect of fungal content on Heterodera schachtii control

Previous laboratory studies have shown that part of the fungus dies during the process of drying. For this reason, not only moist capsules with a fungal content of 1% but also capsules with a fungal content of 10% were produced. The efficacy of dry capsules containing 1% and 10% *H. rhossiliensis* was tested against *H. schachtii* in two 300 ml pot trials. Both trials were conducted with heat-treated field soil, which was mixed with sand at a ratio of 2:1 (w/w) and the pH of the substrate was 6. The first trial was conducted with capsules made of the biopolymer Guargum MF+PA5. Pectine derivate PA5 was added to the Guargum MF capsules as a stabilizer and to prevent capsules from sticking together during the process of drying. The second trial was conducted with capsules made of the biopolymer Guargum MG. The capsules were dried overnight on a clean bench. Remaining capsule moisture was 7.5%. Vitality tests were conducted parallel to pot trials to determine whether or not capsules were contaminated by other organisms.

Each trial consisted of the following treatments:

(1) Non-treated control (Con)
(2) Fungal biomass of *H. rhossiliensis* (H.r.)
(3) Capsules containing 1% *H. rhossiliensis* (MF+PA5 or MG, resp.)
(4) Capsules containing 10% *H. rhossiliensis* (MF+PA5 or MG. resp.)

Each treatment had ten replicates. Fungal biomass (0.12 g biomass suspended in 12 ml of water per pot) or dried capsules containing *H. rhossiliensis* (0.9 g per pot) respectively were mixed thoroughly into the soil. Seven days later, three-day-old sugar beet seedlings were planted and inoculated with 3000 infective J2 of *H. schachtii* in 3.6 ml water. The pots were set in a greenhouse at 20°C ± 3°C. The experiments were set up in a randomized block design. Water was applied as needed. During the trial, plants were fertilized twice with the liquid fertilizer Wuxal Super[®] (0.05%). After eight weeks one nematode generation was completed and the trial was evaluated. Plant shoot fresh weights (g) were determined and cysts were extracted from soil by means of centrifugation according to Caveness and Jenson (1955). The numbers of cysts as well as cyst contents (eggs and juveniles) were determined.

55

5.3 Results

5.3.1 Comparison of Guargum MG, Guargum MF and alginate as capsule material

Capsules types made of Alginate had a diameter of 1.16-1.51 mm. Capsules made of the biopolymer Guargum MG had a capsule diameter of 1.46-1.53 mm. Capsules of the biopolymer Guargum MF were larger (1.62-2.08 mm) and had an elliptic flat form (Figure 14).

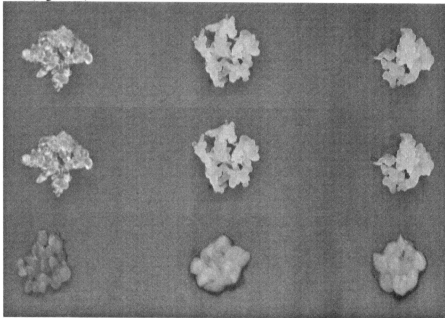

Figure 14: Moist capsules after formulation; from top to bottom: Alginate capsules, Guargum MG capsules, Guargum MF capsules; from left to right: empty capsules, capsules with 3% baker's yeast and capsules with 3% baker's yeast and 1% H. rhossiliensis.

The vitality test confirmed that capsules contained *H. rhossiliensis* as only organism. Aerial mycelia of *H. rhossiliensis* grew successfully from alginate capsules and Guargum MG capsules within two days and from Guargum MF capsules within four days.

In **heat-treated soil**, the application of moist Guargum MG capsules (MG+BY+H.r.) and Guargum MF capsules (MF+BY+H.r.) each containing *H. rhossiliensis* led to a significant reduction of nematode invasion in sugar beet seedlings compared with the non-treated control (CON) by 91% and 55%, respectively. Application of *H. rhossiliensis* as a fungal suspension reduced root penetration by 55%. In contrast, alginate capsules containing *H. rhossiliensis* (ALG+BY+H.r.) had no effect on nematode penetration. Application of all three capsule materials alone without *H. rhossiliensis* increased root penetration by *H. schachtii* significantly in comparison to the non-treated control (Figure 15). The different capsule types did not affect seed germination or plant growth during the trial (data not shown).

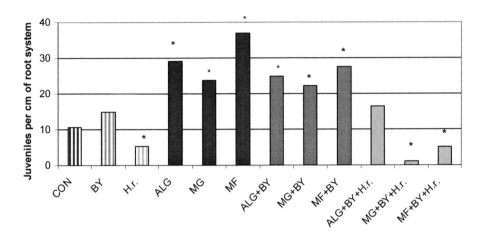

*Figure 15: Effect of moist capsule types applied in **heat-treated soil** on invasion of Heterodera schachtii in sugar beet roots. CON = Non-treated control, BY = Autoclaved baker's yeast, H.r. = Biomass of Hirsutella rhossiliensis as a liquid suspension, ALG = Alginate capsules, MG = Guargum MG capsules, MF = Guargum MF capsules, ALG+BY = Alginate capsules containing baker's yeast, MG+BY = Guargum MG capsules containing baker's yeast, MF+BY = Guargum MF capsules containing baker's yeast, ALG+BY+H.r. = Alginate capsules containing baker's yeast and H. rhossiliensis, MG+BY+H.r. = Guargum MG capsules containing baker's yeast and H. rhossiliensis; means, * = significantly different to control according to Tukey HSD-Test with p ≤ 0.05, n = 10.*

If applied to **non-treated soil**, all of the treatments whether including *H. rhossiliensis* or not, with exception of fungal biomass of *H. rhossiliensis* as a liquid suspension (H.r.), led to an incline of nematode penetration into the roots of sugar beet seedlings. Not all the differences were significant to the non-treated control (CON) (Figure 16). Although a treatment of *H. rhossiliensis* as a liquid suspension did not increase root invasion, penetration of juveniles of *H. schachtii* into the roots was not reduced in comparison to the non-treated control. Germination of sugar beets as well as plant growth was not affected by the capsules.

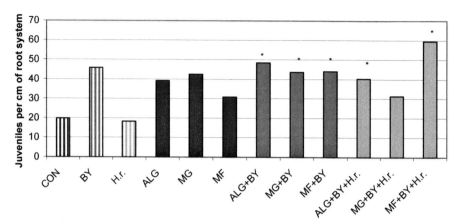

Figure 16: Effect of moist capsule types applied in **non-treated soil** *on invasion of* Heterodera schachtii *in of sugar beet roots. CON = Non-treated control, BY = Autoclaved baker's yeast, H.r. = Biomass of* Hirsutella rhossiliensis *as a liquid suspension, ALG = Alginate capsules, MG = Guargum MG capsules, MF = Guargum MF capsules, ALG+BY = Alginate capsules containing baker's yeast, MG+BY = Guargum MG capsules containing baker's yeast, MF+BY = Guargum MF capsules containing baker's yeast, ALG+BY+H.r. = Alginate capsules containing baker's yeast and H. rhossiliensis, MG+BY+H.r. = Guargum MG capsules containing baker's yeast and H.* rhossiliensis; *means, * = significantly different to control according to Tukey HSD-Test with p ≤ 0.05, n = 10.*

5.3.2 Effect of fungal content on Heterodera schachtii control

The average diameter of dried Guargum MF+PA5 capsules containing *H. rhossiliensis* was 1.05 mm for capsules with 1% fungal biomass and 1.17 mm for capsules with 10% biomass of *H. rhossiliensis*. Capsules of Guargum MG, that were dried, clumped together and were crushed with mortar and pestle prior to use (Figure 17). Subsequently, the average diameter of Guargum MG capsules was 0.66 mm for capsules containing 1% *H. rhossiliensis* and 0.77 mm for capsules containing 10% *H. rhossiliensis*.

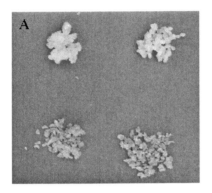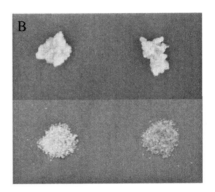

Figure 17: A) Guargum MF+PA5 capsules containing Hirsutella rhossiliensis, *from left to right: capsules with 1% and 10% biomass of* H. rhossiliensis; *top: moist, bottom: dried; B) Guargum MG capsules with* H. rhossiliensis, *from left to right: capsules with 1% and 10% biomass of H. rhossiliensis; top: moist, bottom: dried.*

Vitality assays confirmed that all capsules containing *H. rhossiliensis* were not infested by other organisms prior to trial set-up. Capsules regained two thirds of their actual size within 24 hours if placed on moist filter paper. After three days, growth of aerial mycelia from Guargum MF+PA5 capsules containing *H. rhossiliensis* was microscopically visible. Growth of aerial mycelia from Guargum MG capsules containing *H. rhossiliensis* was delayed. After three days, the fungus grew out of two of ten capsules and only after eight days out of nine of ten capsules.

Guargum MF+PA5

The application of dried Guargum MF+PA5 capsules containing 10% biomass of *H. rhossiliensis* led to a higher shoot fresh weight of 3.63 g in comparison to the non-treated control (CON) with 2.25 g. An application of Guargum MF+PA5 capsules with 1% biomass of *H. rhossiliensis* (MF+PA5+1% H.r.) increased upper shoot fresh weights by approximately 24% in comparison to the control, however differences were not significant. An application of *H. rhossiliensis* as a suspension did not influence growth of plant shoot fresh weights if compared to the non-treated control (Figure 18).

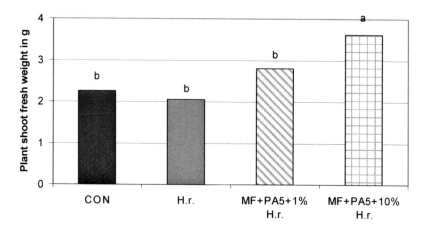

Figure 18: Average plant shoot fresh weights of sugar beet ›Penta‹ (g) eight weeks after applica-tion of dried Guargum MF+PA5 capsules containing Hirsutella rhossiliensis *in heat-treated field soil. CON = Non-treated control, H.r. = Biomass of* H. rhossiliensis *as a liquid suspension, MF+PA5 = Guargum MF+PA5 capsules with 1% or 10% biomass of* H. rhossiliensis, *respec-tively. Means with the same letter are not significantly different according to Tukey HSD with p ≤0.05, n = 10.*

The application of Guargum MF+PA5 capsules containing 1% and 10% *H. rhos-siliensis* led to a significant reduction in the production of eggs and juveniles of *H. schachtii* of 86% and 90%, respectively. In contrast, an application of *H. rhossilien-sis* as a fungal suspension did not influence nematode reproduction.

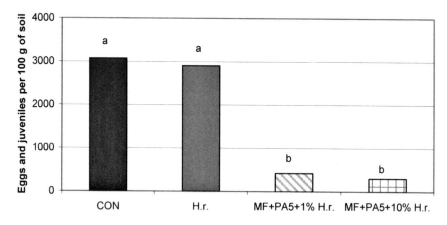

Figure 19: Number of eggs and juveniles of Heterodera schachtii *produced per 100 g of soil after application of dried Guragum MF+PA5 capsules containing* Hirsutella rhossiliensis. *CON = Non-treated control, H.r. = Biomass of* H. rhossiliensis *as a liquid suspension, MF+PA5 = Guargum MF+PA5 capsules containing 1% or 10% biomass of* H. rhossiliensis., *Means with the same letter are not significantly different according to Tukey HSD with p ≤ 0.05, n = 10.*

Guargum MG

The application of Guargum MG capsules containing 1% and 10% *H. rhossilien-sis* as well as a suspension of this fungus led to a significant increase of upper plant weight in comparison to the non-treated control. The highest shoot fresh weight of 5.7 g was achieved after an application with Guargum MG capsules containing 10% *H. rhossiliensis*, followed by Guargum MG capsules containing 1% *H. rhossiliensis* (4.4 g) and *H. rhossiliensis* as a fungal suspension (3.6 g) (Figure 20).

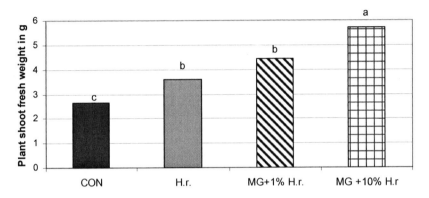

Figure 20: Average shoot fresh weights of sugar beet ›Penta‹ (g) eight weeks after application of dried Guargum MG capsules containing fungus in heat-treated field soil. CON = Non-treated control, H.r. = Biomass of Hirsutella rhossiliensis *as a liquid suspension, MG = Guargum MG capsules containing 1% or 10% biomass of* H. rhossiliensis, *Means with the same letter are not significantly different according to Tukey HSD with p ≤ 0.05, n = 10.*

Treatments of dried Guargum MG capsules containing 1% and 10% biomass of *H. rhossiliensis* led to a significant reduction of nematode reproduction of 75% or 64%, respectively. An application of fungal suspension did not control *H. schachtii* and even led to an increase in nematode reproduction (3736 E+J/100 ml soil) in comparison to the control (2460 E+J/100 ml soil) (Figure 21).

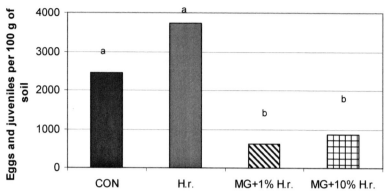

Figure 21: Number of eggs and juveniles of Heterodera schachtii *produced per 100 g of soil after application of dried Guargum MG capsules containing Hirsutella rhossiliensis. CON = Non-treated control, H.r. = Hirsutella rhossiliensis as a liquid suspension, MG = Guargum MG capsules containing 1% or 10% biomass of* H. rhossiliensis., *Means with the same letter are not significantly different according to Tukey HSD with p ≤ 0.05, n = 10.*

5.4 Discussion

The endoparasitic fungus *H. rhossiliensis* is a widespread parasite of plant-parasitic nematodes (Jaffee and Muldoon, 1989; Jaffee et al., 1989; Sturhan and Schneider, 1980). Since its discovery in 1980, many research groups have focused on the development of this fungus as a biological control agent (Jaffee et al. 1996; Viaene and Abawi, 2000; Liu and Chen, 2005). Application of the fungus as a mycelial suspension can lead to nematode controls (Lackey et al., 1992; Liu and Chen, 2005), but is not suitable for practical use because shredded vegetative hyphae in a water suspension is not homogenous in distribution within a tank and the suspension can clog nozzles of sprayers. Furthermore, vegetative hyphae of *H. rhossiliensis* do not have a long shelf-life. A suitable formulation is required for the development of *H. rhossililiensis* as biological control agent, which guarantees a safe and easy application of the product. As the fungus is a weak competitor, the formulation should further ensure a good establishment of the fungus in the soil in order to obtain a high level of parasitized plant-parasitic nematodes. Alginate formulations were favored by many research groups but the capsules did not prove to be effective in the field (Jaffee et al., 1996; Lackey et al., 1993; Jaffee and Muldoon, 1997). Investigations by Patel (1998), Gutberlet (2000) and Rose (2000) of fungal capsules based on cellulose showed promising results under field conditions. Following on these results, experiments were conducted with Guargum derivates as encapsulation material for *H. rhossiliensis.*

The results of the experiments showed that an application of capsules made of different Guargum derivates can reduce nematode infestation of sugar beets up to 90% in heat-treated soil. In the biotest conducted with moist capsules made of Guargum

MF and Guargum MG containing 1% *H. rhossiliensis*, nematode infestation was reduced significantly in comparison to the non-treated control. However, capsules made of alginate containing *H. rhossiliensis* did not reduce root infestation with *H. schachtii*, which agrees with observations made by Jaffee et al. (1996) on cabbage. Furthermore, the three capsule types without *H. rhossiliensis* showed an adverse effect and promoted juvenile penetration into the roots. The reason for this cannot be fully explained. Studies by Lackey et al. (1993) have shown that low application rates of alginate capsules without fungus did not have an affect on nematode invasion or plant growth, respectively. Perhaps our application rate was too high. Schuster and Sikora (1992) stated that high concentrations of alginate per unit soil may suppress plant growth and nematode development but this was not evident in our experiments as the germination rate of sugar beet was over 80% plant growth was not inhibited in all of the experiments conducted nor was the development of nematodes hemmed.

The next step consisted of drying the capsules. Dried capsules are generally preferred not only by the growers, but also by the producer, trading companies, etc. as they are easy to store and to apply. Dried capsules of Guargum MF+PA5 and Guargum MG containing *H. rhossiliensis* had a positive effect on plant growth, although plant growth promotion was not significant in all cases. It can be stated that an increase of the fungal content in the capsules from 1% to 10% led to a significant increase in plant shoot fresh weight but it did not have a significant effect on the nematode population density in the soil in comparison to a capsule treatment with 1% *H. rhossiliensis*. As all other factors were identical, perhaps an increase of fungal biomass in the capsules stimulated plant growth directly or otherwise indirectly through the release of nutrients after microbial decomposition of fungal mycelia. To clarify this assumption in future tests, treatments with 10% fungal biomass as a liquid suspension and capsules without biomass have to be added. Dried capsules of Guargum MF+PA5 or Guargum MG containing *H. rhossiliensis* did not only promote plant shoot fresh weight but also significantly reduced the nematode population density up to 90% after 8 weeks. The reduction of the nematode population density was comparable to the results obtained in the biotest with moist capsules containing *H. rhossiliensis*. Although MG capsules containing *H. rhossiliensis* were crushed after they were dried, it did not seem to have a negative effect on the capsule's efficacy.

Costs play a major role in the development of a biological control agent and the amount of active ingredient (fungus concentration) contributes significantly to it. The content has to be high enough to secure good parasitism of nematodes in the field, however, if too high, it could have a negative impact on the efficacy of the formulation. The results of the pot trials showed that increasing the fungal content from 1% to 10% led to an increase of shoot fresh weight but it did not improve the efficacy against *H. schachtii*. Perhaps, the constant factor »baker's yeast« at 3% did not satisfy the demand the additional amount of fungus (1%→10%) in the capsule evoked. Research must be conducted to determine in which relation nutrient and fungal components should be added to the capsule for optimal growth of aerial mycelia of *H. rhossiliensis*.

Although moist and dried capsule formulations showed a good efficacy in heat-treated soil, results of moist capsules were dissatisfying in non-treated soil. The biotest in non-treated soil showed that an application of capsules with and without fungus led to an increase of root penetration of up to 300% in comparison to the control. An application of *H. rhossiliensis* as fungal biomass did not increase or reduce nematode invasion compared to the control. The reasons for an increase in nematode invasion in the root system after an application of capsules with and without fungus could indicate a lack of growth of *H. rhossiliensis* in the soil or a side-effect in connection with the deterioration of organic substance in the soil. Low vitality of the fungus, the contamination of capsules or too low amount of *H. rhossiliensis* could be excluded as possible reasons. Jaffee et al. (1996) also reported that alginate capsules containing *H. rhossiliensis* did not have a good effect against *H. schachtii* in non-treated soil. As *H. rhossiliensis* is a weak competitor, they assumed that fungal growth was suppressed by other antagonistic organisms. This could explain our own results as the capsule material itself or nutrients in the capsule might act as a nutrient source for microorganisms. Not only does *H. rhossiliensis* have to compete against an increased population density of microorganisms surrounding the capsule material, but it must also produce phialides and conidia in soil to infect and parasitize plant-parastic nematodes. Research on soil applications of 1% chitin have shown that bacteria populations increase 100 fold within 24 hours and bacteria populations increase 50 fold (Hallmann et al., 1999). Perhaps the hurdle is too high for the slow growing and weak competitor *H. rhossiliensis*.

5.5 Summary

In the present study, research was conducted to determine the efficacy of various types of encapsulation of the endoparasitic fungus *Hirsutella rhossiliensis* to control the sugar beet cyst nematode *Heterodera schachtii* on sugar beet. The effectiveness of the moist capsule types Guargum MF and Guargum MG containing 1% *H. rhossiliensis*, was investigated in 100 ml bioassays which were conducted in heat-treated and non-treated soil. The number of nematodes in each root system was counted seven days after germination. The efficacy of dried capsules that were made of the biopolymers Guargum MF + Pectine PA5 and Guargum MG, containing 1% and 10% *H. rhossiliensis*, were tested for their effectiveness against *H. schachtii* in two separate pot trials (300 ml) on sugar beet. The two pot trials were conducted in heat-treated soil. Results of the bioassays showed that all capsule types containing *H. rhossiliensis* led to a significant reduction of nematode invasion up to 90% when applied to heat-treated soil. Increasing the amount of fungus added to a capsule from 1% to 10% did not improve its efficacy. In contrast, the capsule materials stimulated root penetration by *H. schachtii*. When applied to non-treated soil, capsules containing *H. rhossiliensis* failed to suppress nematode invasion of roots of sugar beet seedlings. In fact, in some cases it even led to an increase in root penetration. As *H. rhossiliensis* is an inferior competitor and its growth is quite slow, it can be assumed, that its growth was inhibited by other soil microorganisms. An encapsulation of *H. rhossiliensis* as described in this study is not suitable for control of *H. schachtii*. Formulations must therefore be further optimized in order to achieve a better biological control affectiveness e. g. modification of the outer pH-level of the capsule to a level not favoured by bacteria or fungi.

6 Study on pH-tolerance of encapsulated *Hirsutella rhossiliensis* and its effect on antagonistic organisms

6.1 Introduction

The formulations of *H. rhossiliensis* with Guargum derivates and alginate were effective in heat-treated soil. However, if applied in non-treated soil, microbial contamination of the capsules suppressed the growth of *H. rhossiliensis*. In order to overcome this problem, the pH of the outer capsule was modified to be suboptimal for microbial contaminants. According to Jaffee and Zehr (1983) and Liu and Chen (2002) the optimum pH for growth and sporulation of *H. rhossiliensis* lies between 5.3 and 7. The optimum range of pH preferred by most soil bacteria is around 6.3-6.8. Thus dropping the outer pH of the capsule to a more acidic level of 3, 4 or 5 might prevent bacteria from multiplying on the capsule surface even if capsules were placed in field soil with a pH that was nearly neutral. On the other hand fungi have a broader pH tolerance and tend to multiply at lower values. A prevention of capsule infestation might be obtained by raising the outer pH-level of the capsule to 7 or 8. In this study the performance of different types of encapsulation of *H. rhossiliensis*, dipped in solutions with pH-levels ranging from 3 to 8 and placed on heat-treated soil, non-treated soil and WA, were compared. Unfortunately, the infestation of capsules by both fungal and bacterial contaminants cannot be prevented simultaneously by a modification of the outer pH-level of the capsule. The broad spectrum chosen within the pH range was investigated to determine which type of contaminant had a greater impact on growth of *H. rhossiliensis*. Furthermore, the effect of a modification of the pH-level of the outer capsule surface to a level not optimal for the growth of *H. rhossiliensis* was investigated.

6.2 Material and Methods

6.2.1 Experimental design

Two *in vitro* experiments were conducted with different capsule formulations containing *H. rhossiliensis*. The capsules were produced and delivered by BIOCARE GmbH. Capsule types Guargum MF+PA5 and Guargum MG+PA2 had an elliptic form, whereas capsule types Guargum MG and Alginate ALG+PA5 were round. The first *in vitro* experiment was conducted with pH solutions that did not contain a buffer, whereas the second experiment was conducted with pH solutions that contained a citrate-phosphate buffer. The pH solution was prepared by filling one liter glass bottles with 800 ml of solution and adjusting the pH to 3, 4, 5, 6, 7 or 8. The pH solution was then sterilized by autoclaving it at 120°C for 45 minutes. At the beginning of each experiment, forty capsules of each capsule type were incubated in the different pH solutions. Therefore, sterile 250 ml glass beakers were filled with 100 ml of a pH-solution to which the capsules were then added. The beakers containing a pH solution and a set of forty capsules were covered with aluminium foil to prevent contamination and set on a rotary shaker for ten minutes at 225 rpm to keep the capsules from sinking to the bottom of the beaker. Hereafter, the capsules were removed from the pH solution. Ten treated capsules were placed in a Petri plate filled with 45 g of either non-treated field soil or heat-treated field soil or on WA. Heat-treated field soil had previously been heated for two hours at 180°C. Both soil types had a soil moisture content of 6% and a pH $CaCl_2$ of 6.2, whereas WA had a pH of 6.7. The Petri plates were placed in an incubator at 23°C. Every two to three days, radial growth of aerial mycelia of *H. rhossiliensis* was measured in mm. The initial capsule diameter was assessed of each capsule at trial set-up and substracted from the diameter of the capsule containing aerial mycelia of *H. rhossiliensis* during the trial. The difference was divided by two to determine the radius of aerial mycelia.

6.2.2 In vitro *test A*

The first *in vitro* experiment was conducted with the following four capsule types: Alginate ALG+PA5, Guargum MF+PA5, Guargum MG and Guargum MG+PA2. Each capsule type contained 5% *H. rhossiliensis*. The capsules were incubated in solutions, based on autoclaved water that were adjusted to the pH of 3, 4, 5, 6, 7 or 8 by adding HCL 1 mol or NaOH 0.02 mol until the desired pH was reached. The experiment was conducted as described above in 6.2.1 Experimental design.

6.2.3 In vitro *test B*

In vitro test A showed that, when placed on WA, capsule types Guargum ALG+PA5 and MG+PA2 containing 5% *H. rhossiliensis* proved best radial growth of aerial mycelia of *H. rhossiliensis*. For this reason these two capsule types were chosen for additional testing in a buffered pH system. The citrate-phosphate buffer was prepared according to Bast (2001) and adjusted to pH 3, 4, 5, 6 and 7. One-hundred ml of the pH solutions were filled into 250 ml glass beakers set on a magnet stirring plate and adjusted to the exact pH-level by adding one of the two buffer component solutions. The solutions were then autoclaved at 120°C for 45 minutes prior to trial set-up. Capsules were added to the mixtures for ten minutes and afterwards placed in Petri plates filled with heat-treated soil, non-treated soil or 1.5% WA as described in chapter 6.2.1.

6.3 Results

6.3.1 In vitro *test A*

Soaking capsules in an aquatic solution with a pH adjusted to 3, 4, 5, 6, 7 or 8 did not affect the growth of aerial mycelia of *H. rhossiliensis* from the capsules regardless of the capsule type and the pH level if capsules were placed on WA. Statistical results proved that there were interactions between the material used to encapsulate

H. rhossiliensis and the pH solution the capsules were soaked in. Furthermore, significant differences of fungal growth were detected between the individual pH levels and also between the selected encapsulation materials. When comparing the four different encapsulation materials, radial growth of aerial mycelia was best from Alginate ALG+PA5 capsules (1.63 mm to 1.93 mm) followed by Guargum MG+PA2 containing *H. rhossiliensis* (1.36 mm to 1.87) (Figure 22). Adding pectine derivate PA2 to the formulation significantly improved the performance of *H. rhossiliensis*, if results were compared to growth of aerial mycelia from Guargum MG capsules (0.85 mm to 1.32 mm). Even radial growth of aerial mycelia from MF+PA5 capsules (1.11 mm to 1.52 mm) was better than that from MG capsules. Not only did encapsulation material have an affect on radial growth but also the pH solution, in which the capsules had been soaked, showed an effect. Radial growth from Alginate Alg+PA5 capsules was not affected by a pH solution. Fungal growth from MF+PA5 and MG+PA2 capsules was better at the pH levels of 3, 4, and 5 and declined at pH levels 6, 7 and 8. Radial growth from Guargum MG capsules was worse at the highest and lowest pH level (3, 8) (Figure 22).

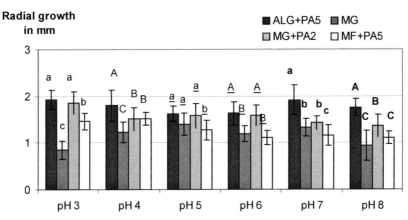

Figure 22: Radial growth of aerial mycelia of Hirsutella rhossiliensis *in mm fifteen days after trial set-up from capsule types Alginate ALG+PA5 (ALG+PA5), Guargum MG (MG), Guargum MG+PA2 (MG+PA2) and Guargum MF+PA5 (MF+PA5) containing 5% H. rhossiliensis that had been soaked in a bath with a pH of 3, 4, 5, 6, 7 and 8 for ten minutes and had been placed on WA. Means ± standard deviation with the same letter are not significantly different based on Holm-Sidak (All Pairwise Multiple Comparison Procedures), P ≤ 0.05; n = 10.*

If capsules were placed on heat-treated soil, aerial mycelia of *H. rhossiliensis* visibly grew out of the capsules, regardless of the pH-solution the capsules had been dipped in or the capsule type (Figure 23). Since the capsule itself shrunk, radial growth of aerial mycelia could not be measured even though growth of aerial mycelia was visible. It was not possible to establish a positive difference in size. Capsules remained intact over the duration of the trial, however, those placed on untreated soil slowly disintegrated. Capsules, placed on non-treated soil, were infested by antagonistic organisms in spite of the pH solution the capsules had been dipped in (Figure 23). Within two weeks the entire capsule had deteriorated.

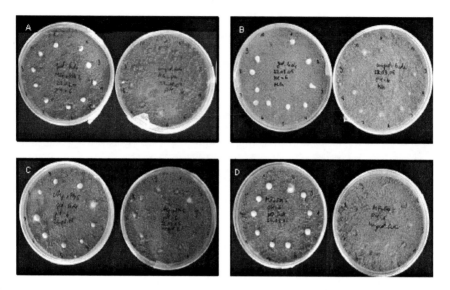

Figure 23: The pictures display capsules thirteen days after trial set-up that had previously been dipped in an aquatic solution with a pH of 6 for ten minutes and placed in Petri dishes filled with heat-treated (left) and non-treated soil (right). Growth of aerial mycelia of Hirsutella rhossiliensis *can be observed from capsules placed on heat-treated soil. Capsules placed on non-treated soil were infested by other organisms and have almost diminished: (A) Ten Guargum MG + Pectine derivate PA2 capsules placed on heat-treated soil (left) and non-treated soil (right); (B) Ten Guargum MG capsules placed on heat-treated soil (left) and non-treated soil (right); (C) Ten Alginate ALG + Pectine derivate PA5 capsules placed on heat-treated soil (left) and non-treated soil (right); (D) Ten Guargum MF + Pectine derivate PA5 capsules placed on heat-treated soil (left) and non-treated soil (right).*

6.3.2 In vitro *test B*

Growth of *H. rhossiliensis* from capsule types Alginate ALG+PA5 and Guargum MG+PA2 was not inhibited, when soaked in a citrate-phosphate buffer with the pH adjusted to 3, 4, 5, 6 or 7 for ten minutes and placed on WA. Alginate ALG+PA5-capsules, soaked in citrate-phosphate buffer pH 6 and 7, started to dissolve after several minutes, but could still be recovered and were placed on WA after ten minutes. This capsule type was not stable in citrate-phosphate buffer, whereas Guargum MG+PA2 did not show any signs of instability. When comparing the growth of *H. rhossiliensis* from both capsule types at different pH levels, they did not significantly differ at a pH of 3, 4 and 5. At pH 6, radial growth of *H. rhossiliensis* was significantly worse from Guargum MG+PA2 than from Alginate ALG+PA5 capsules and at a pH of 7 it was the opposite (Figure 24). Radial growth of aerial mycelia from ALG+PA5 capsules ranged from 1.63 mm to 2.01 mm. Radial growth of aerial mycelia from MG+PA2 capsules ranged from 1.25 mm to 1.98 mm.

If the capsules were placed on heat-treated soil, growth of aerial mycelia of *H. rhossiliensis* could only be observed microscopically but not visually. Furthermore, approximately 20% of the capsules regardless of the capsule type were infested with a different type of fungus. If capsules were placed on non-treated soil, all became infested by other microorganisms regardless of the pH level of the solution the capsules had been dipped in (Figure 25). Capsules placed on non-treated soil started to deteriorate shortly after becoming infested by other microorganisms whereas capsules placed on heat-treated soil remained intact over the duration of the trial.

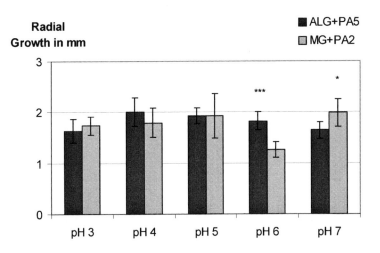

Figure 24: Radial growth of aerial mycelia of Hirsutella rhossiliensis *in mm thirteen days after trial set-up from capsule types Alginate ALG+PA5 (ALG+PA5) and Guargum MG+PA2 (MG+PA2) containing 5% H. rhossiliensis that had been soaked in a citrate-phosphate buffer solution with a pH of 3, 4, 5, 6, and 7 for ten minutes and had been placed on WA. * = significantly different to control based on Holm-Sidak (All Pairwise Multiple Comparison Procedures), P ≤ 0.05, *** = significantly different to control based on Holm-Sidak (All Pairwise Multiple Comparison Procedures), P ≤ 0.01, n = 10.*

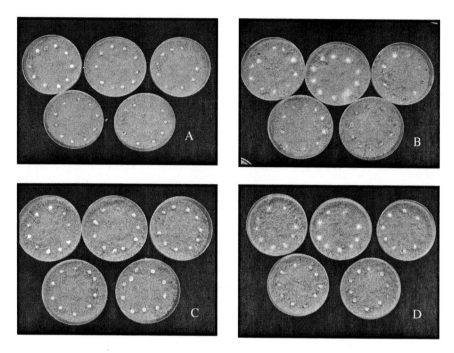

Figure 25: Display of ten capsules containing 5% Hirsutella rhossiliensis placed on heat-treated soil (A, C) or non-treated soil (B, D) in each of five Petri dishes ten days after trial set-up; (A, B) Capsule type Alginate ALG + Pectine derivate PA5; (C, D) Capsule type Guargum MG + Pectine derivate PA2.

6.4 Discussion

In both *in vitro* tests, if the capsules were placed on WA, aerial mycelia of *H. rhossiliensis* grew from different encapsulation materials whether capsules had been dipped in water or a citrate-phosphate buffer solution with an adjusted pH of 3, 4, 5, 6, 7, or 8. Results of *in vitro* test A showed that encapsulation material had a stronger affect on fungal growth than the pH of the capsule. Best radial growth of aerial mycelia from capsules was observed from Alginate ALG+PA5 capsules and Guargum MG+PA2 capsules placed on WA. Adding the component pectine derivate PA2 to the formulation Guargum MG significantly improved radial growth if compared to growth from Guargum MG capsules without PA2. In fact, all formulations containing a pectine derivate suited better as encapsulation material for *H. rhossiliensis* than Guargum MG without this component. Furthermore, results of both *in vitro* tests showed that the solution with an adjusted pH, in which the capsules had been soaked for ten minutes, also had an effect on fungal growth. According to Jaffee and Zehr, 1983 and Liu and Chen, 2002, the optimum pH for growth and sporulation of *H. rhossiliensis* lies between 5.3 and 7. In our studies we had hypothesized that soaking the capsules in a solution with a pH not optimal for *H. rhossiliensis* might affect the growth of the fungus. Contrary results were obtained in *in vitro* test A as growth of *H. rhossiliensis* from capsule types Guargum MG+PA5 and Guargum MF+PA5 was better at lower pH levels of 3, 4 and 5 and became worse at more neutral pH levels. Nonetheless, growth of *H. rhossiliensis* from Alginate Alg+PA5 was best, thus underlining the fact that the encapsulation material had the strongest affect on fungal growth. Interactions between the encapsulation material and pH, the capsules were soaked in, were found. If soaked in water with an adjusted pH level between 3 and 8, growth of *H. rhossiliensis* from Alginate Alg+PA5 did not differ significantly between the pH levels. When soaking the capsules in a citrate-buffer solution with a pH level between 3 and 7, radial growth of *H. rhossiliensis* was significantly worse at the highest and lowest pH level. In order to confirm results, that the factors pH solution and the encapsulation material interacted with each other and that each factor had its own influence on fungal growth, the *in vitro* experiments have to be repeated. More importantly though, we can conclude that both pH solutions did not inhibit growth of *H. rhossiliensis*.

Although radial growth could be measured if capsules were placed on WA, it could not be recorded if capsules were placed on heat-treated soil. *In vitro* test A showed that, if capsules were soaked in water with an adjusted pH-level and placed on heat-treated soil, aerial mycelia grew visibly from the capsule but the capsule itself started to shrink so a difference in size could not be established. In *in vitro* test B, the capsule types ALG+PA5 and MG+PA2 exhibited poor growth of aerial mycelia that could only be observed microscopically. Nonetheless, capsules remained intact in both experiments, an indication that capsules had not been infested by other organisms. Previous studies also reported that if soil is heat-treated, organisms may be suppressed that otherwise induce dormancy (Jaffee and Zasoski, 2001; Liebman and Epstein, 1992; Lockwood, 1988). In retrospect, perhaps soil moisture of 6% was not optimal for *H. rhossiliensis*. Later trials showed that raising soil moisture to 13% improved radial growth.

If placed on non-treated soil, *H. rhossiliensis* did not grow from any capsule type regardless of the pH solution the capsules had been soaked in. In this case either *H. rhossiliensis* was very sensitive towards soil fungistasis and remained dormant in non-treated soil or the fungus was killed by another microorganism. It was not possible to isolate *H. rhossiliensis*. Capsules were infested by other microorganisms and deteriorated quickly. As field soil had a pH of 6.2, perhaps it was not possible to maintain the established pH level on the outer surface of the capsule. If the entire capsule had its pH adjusted to a certain level, perhaps antagonistic organisms would not have been able to suppress *H. rhossiliensis*. The capsules were dipped in a solution with molecules that are permeable and can wander through the cells and back out again easily, making the outer pH level very instabile. Jaffee (1999) found that pelletized *H. rhossiliensis* did well in a commercial vineyard whether the soil was heat-treated or not. The pellets performed well in the non-treated field soil of the commercial vineyard, where the soil pH was quite acidic. He inferred that soil pH might explain why the formulation was more active in a vineyard with acidic soil than in a vineyard with a soil pH that was nearly neutral. He assumed that low soil pH might reduce the abundance of bacteria in field soil. Our results showed that by lowering the pH of the outer layer of capsules, the formulation was not improved and could not hinder the suppression by antagonistic organisms as well as heat-treated soil or low soil pH might.

6.5 Summary

Previous experiments showed that encapsulation of *H. rhossiliensis* only worked in heat-treated soil, but not in non-treated soil. To make encapsulation more resistant to microbial inhibition, the capsules were soaked in water and citrate-phosphate solutions with adjusted pH-levels of 3, 4, 5, 6, 7 or 8 in order to change the pH of the outer capsule layer. Acidic pH levels 3, 4 or 5 were chosen to prevent bacteria from multiplying on the capsule surface, neutral pH-levels 7 and 8 were chosen to hinder fungal contamination. Treated capsules were placed on 1.5% WA, heat-treated soil and non-treated soil. The effect of pH solutions on growth of aerial mycelia of *H. rhossiliensis* was investigated. Furthermore, fungal growth from different encapsulation materials was compared. Results of *in vitro* test A and B showed that the components of either pH solutions did not inhibit fungal growth from different capsule types if capsules were placed on WA. Capsule material had the greatest influence on growth of aerial mycelia. Best radial growth of aerial mycelia was observed from Guargum MG+PA2 capsules and Alginate ALG+PA5 capsules containing *H. rhossiliensis*. If capsules were placed on heat-treated soil, they were not contaminated by antagonistic organisms regardless of the pH solution they had been soaked in and radial growth of aerial mycelia of *H. rhossiliensis* was observed. If capsules were placed on non-treated soil, *H. rhossiliensis* did not grow from any capsule type regardless of the pH solution the capsules had been soaked in. Capsules were infested by microbial organisms and growth of *H. rhossiliensis* from the capsules was inhibited. A modification of the pH level of the capsule did not improve the formulation. Contamination of capsules by other microorganisms could not be prevented.

7 Studies on the efficacy of Alginate ALG+PA5 as an encapsulation material for *Hirsutella rhossiliensis*

7.1 Introduction

The endoparasitic fungus *H. rhossiliensis* is capable of naturally suppressing plant-parasitic nematodes (Eayre et al., 1987, Jaffee and Muldoon, 1989, Jaffee and Zehr, 1982, Müller, 1982). Furthermore, following artificial application, the fungus significantly suppressed plant-parasitic nematodes (Velvis and Kamp, 1996 and Amin, 2000). Despite these promising results, *H. rhossiliensis* is not yet available as a biocontrol agent, mainly due to its inconsistent efficacy and non-availability of an acceptable formulation. The research of this thesis focused on the efficacy of different types of encapsulation of *H. rhossiliensis* against plant-parasitic nematodes. In chapter 6 »Study on pH-tolerance of encapsulated *Hirsutella rhossiliensis* and its effect on antagonistic organisms«, radial growth of *H. rhossiliensis* was examined for four different capsule types. The capsule type Alginate ALG+PA5 achieved best results under sterile conditions and was further studied. In a first experiment, the importance of soil type on growth of the fungus *H. rhossiliensis* from Alginate ALG+PA5 capsules containing *H. rhossiliensis* plus baker's yeast as a nutrient additive was investigated *in vitro*. Furthermore, in addition to Alginate ALG+PA5 capsules containing *H. rhossiliensis*, assimilative hyphae in form of pure fungal pellets, obtained from liquid culture were included in the study to determine whether or not the capsule material itself might affect the growth of *H. rhossiliensis*. The fungus is a weak saprophyte (Jaffee and Zehr, 1985), therefore one might argue that supplemental nutrients in an encapsulation would enhance fungal sporulation or on the other hand attract soil saprophytes that might use the nutrients and inhibit growth of *H. rhossiliensis*. In regard to this aspect, greenhouse trials were performed in heat-treated soil to eliminate important biotic factors such as fungi and bacteria that influenced growth of the fungus from capsules negatively in previous greenhouse trials. This was also an approach to achieve a consistency in performance in the following studies conducted in heat-treated soil. In the following studies an encapsulation of *H. rhossiliensis* in Alginate ALG+PA5 was tested against the sugar beet cyst nematode *Heterodera schachtii* on sugar beet ›Dorena‹ and *Meloidogyne incognita* on cucumber ›Belcanto F1‹.

7.2 Material and Methods

7.2.1 Influence of different soil types on growth of Hirsutella rhossiliensis

An *in vitro* test was conducted to determine which type of soil was beneficial for the growth of aerial mycelia of *H. rhossiliensis* from alginate ALG+PA5 capsules.

Vegetative, assimilative hyphae, harvested in form of fungal pellets from liquid culture, were used for the production of capsules containing *H. rhossiliensis*. The company BIOCARE produced and delivered moist alginate ALG+PA5 capsules containing 5% *H. rhossiliensis* and 3% baker's yeast, with a diameter of 4.5 to 6 mm. In addition, fungal pellets without encapsulation material were applied in this trial. Fungal pellet sizes ranged from 1 to 5 mm.

Five different soil substrates were chosen for the *in vitro* test:
  (1) silicate sand (<0.70% humus, pH CaCl2 6.2, 2.2% TC (Total Carbon),
  (2) loess (2.7% Humus, pH CaCl2 7.5, 1.6% TC),
  (3) field soil A (1.1% humus, pH CaCl2 6.2, 0.66% TC),
  (4) field soil B (2.9% humus, pH CaCl2 6.5, 1.7% TC),
  (5) field soil C (3.8% humus, pH CaCl2 5.5, 2.2% TC).

The different soil types were non-treated and heat-treated at 180°C for two hours for this experiment. Soil moisture of both heat-treated and non-treated soils were adjusted to 13% prior to trial set-up. Then, 40 g of each soil type, either heat-treated or non-treated were filled into a Petri plate. Ten Alginate ALG+PA5 capsules containing *H. rhossiliensis* or ten fungal pellets were placed on the surface of the soil. Hereafter, the Petri plate was covered with a lid and sealed with parafilm to prevent evaporation. In addition, capsules containing fungus and fungal pellets were placed on 1.5% WA and moist, sterile filter paper in a Petri plate respectively, to check whether or not *H. rhossiliensis* was vital. Each treatment had ten replicates. The Petri plates were placed in an incubator for 15 days at 23°C. Growth of aerial mycelia of *H. rhossiliensis* was examined daily, radial growth of the fungus was measured in mm every four days and fungal growth of capsules and pellets was rated 15 days after trial set-up. For the latter, the rating was based on a scale of 0-5 (0 = absence of fungal growth, 1 = growth of aerial mycelia from 20% of the capsule or pellet surface, visible to the viewer, 2 = 40% visible fungal growth, 3 = 60% visible fungal growth, 4 = 80% visible fungal growth 5 = 100% visible growth).

7.2.2 Efficacy of fungal Alginate ALG+PA5 capsules against Heterodera schachtii *on sugar beet*

The efficacy of Alginate ALG+PA5 capsules containing 5% *H. rhossiliensis* and the individual components of the capsules were investigated against *H. schachtii* on sugar beet ›Dorena‹. Components of the capsules and capsules were delivered by the company BIOCARE.

The treatments of the biotest were:
  (1) Non-treated control (CON)
  (2) Fungal biomass of *H. rhossiliensis* (H.r.)
  (3) Alginate powder (ALG)
  (4) Pectine derivate powder (PA5)
  (5) Alginate ALG+PA5 capsules with 3% baker's yeast (ALG+PA5)

(6) Alginate ALG+PA5 capsules with 3% baker's yeast and 5% *H. rhossiliensis* (ALG+PA5+H.r.)

Product quality was examined in a vitality assay. For this purpose, thirty capsules containing H. rhossiliensis were placed on moist autoclaved filter paper, on PDA and on WA, respectively. Purity of capsules and growth rate of H. rhossiliensis were evaluated. The biotest was conducted in field soil with a humus content of 2.9, a pH CaCl2 of 6.5 and a total carbon content of 1.7%, which was heat-treated for two hours at 180°C in an oven prior to experimental set-up. At trial set-up (day 0), 100 ml of soil were thoroughly mixed with the treatment and filled into a plastic container (Kelder Plastibox b.v., s' Heerenberg, The Netherlands). For the treatment »Fungal biomass of H. rhossiliensis« 0.2 g of vegetative hyphae were suspended in 2 ml of H2O, which were then added to 100 ml of soil. For the treatment Alginate ALG+PA5 capsules with and without 5% H. rhossiliensis, 4 g of moist capsules were added to the soil. As the capsules consisted of 1% alginate powder and 2.5% pectine derivate powder, 0.04 g of alginate powder and 0.1 g of pectine derivate powder were applied to 100 ml of soil. Each treatment had 12 replicates. The containers were set in the greenhouse at an average temperature of $20°C \pm 3°C$. Irrigation was applied as needed. On day 8, the soil of each container was inoculated with 1000 infective second-stage (J2) juveniles of H. schachtii. Also, one sugar beet seed ›Dorena‹ was sown into the centre of each container, approximately 0.5 cm deep. Sugar beet seedlings were extracted from the soil of the containers on day 19. To evaluate the trial, plant shoots were discarded and roots were rinsed free of soil debris under running tap water. Total root fresh weight was determined and the juveniles within the root were stained with a 0.01% acid fuchsin solution (Byrd et al. 1983) and numbers counted under a dissecting microscope.

7.2.3 *Efficacy of fungal Alginate ALG+PA5 capsules against* Meloidogyne incognita *on cucumber*

The biocontrol efficacy of moist Alginate ALG+PA5 capsules was investigated in two biotests against *M. incognita* on cucumber ›Belcanto F1‹.

Treatments were as following:
(1) Non-treated control (CON)
(2) Fungal biomass of *H. rhossiliensis* (H.r.)
(3) Alginate ALG+PA5 capsules with 3% baker's yeast and 5% *H. rhossiliensis* (ALG+PA5+H.r.)

To certify product quality, a vitality assay was conducted. For this purpose, thirty capsules containing H. rhossiliensis were placed on moist autoclaved filter paper, on PDA and on WA, respectively. Purity of capsules and growth rate of *H. rhossiliensis* was determined.

The biotests were conducted in heat-treated and non-treated field soil with a humus content of 2.9, a pH $CaCl_2$ of 6.5 and a total carbon content of 1.7%. At trial set-up (day 0), 100 ml of soil were thoroughly mixed with the treatment and filled into 100 ml plastic containers (Kelder Plastibox b.v., s' Heerenberg, The Netherlands). The treatment »fungal biomass of *H. rhossiliensis*« consisted of 0.2 g of vegetative hyphae suspended in 2 ml of H_2O, which was then added to 100 ml of soil. For the treatment Alginate ALG+PA5 capsules with 5% *H. rhossiliensis*, 4 g of moist capsules were added to 100 ml of soil. Each treatment had 10 replicates. The boxes were set in the greenhouse at an average temperature of 20°C ± 3°C. Irrigation was applied as needed. On day 8, the soil of each container was inoculated with 1000 infective second-stage (J2) juveniles of *M. incognita*. Also, one cucumber seed was sown into each container approximately 0.75 cm deep. Cucumber seedlings were extracted from the soil of the containers on day 18. To evaluate the trial, plant shoots were discarded and roots were rinsed free of soil debris under running tap water. Root fresh was determined in centimetre and juveniles within the root systems were stained with a 0.01% acid fuchsin solution (Byrd et al. 1983) and numbers counted under a dissecting microscope.

7.2.4 Effect of fungal Alginate ALG+PA5 capsules on nematode reproduction of Heterodera schachtii

The efficacy of moist Alginate ALG+PA5 capsules containing 5% *H. rhossiliensis* was tested against *H. schachtii* in two 300 ml pot trials.

The treatments of the pot trials were:
(1) Non-treated control (CON)
(2) Fungal biomass of *H. rhossiliensis* (H.r.)
(3) Alginate ALG+PA5 capsules with 3% baker's yeast (ALG+PA5)
(4) Alginate ALG+PA5 capsules with 3% baker's yeast and 5% *H. rhossiliensis* (ALG+PA5+H.r.)

Product quality was examined in a vitality assay as described previously in »Efficacy of fungal Alginate ALG+PA5 capsules against *Meloidogyne incognita* on cucumber (*Cucumis sativus*)«. The trials were conducted in heat-treated and non-treated field soil, which had been mixed with silicate sand in a ratio of 2:1. The soil substrate had a pH of approximately 6. At trial set-up (day 0), 300 ml of soil were thoroughly mixed with the treatment »fungal biomass of *H. rhossiliensis*« (0.6 g biomass in 10 ml of water/300 ml soil) or moist capsules (12 g/300 ml soil) respectively and 2.4 g of cyst inoculum soil, which contained cysts of *H. schachtii* with a total of approximately 4,500 eggs and juveniles of *H. schachtii* (chapter 2.2.1 Preparation of cyst inoculum of *Heterodera schachtii*). The treated soil was filled into a 300 ml pot. For the treatment »Non-treated control« 10 ml of water were added to the soil before it was filled into the pot. Each treatment had 12 replicates. The pots were set in the greenhouse at an average temperature of 20°C +/- 3°C. Irrigation was applied as needed. On day 12, two sugar beet seeds were sown into each pot. After germination seedlings were thinned to one per pot. Soil temperature was measured as of seed germination on day 15 to determine the completion of a life-cycle of *H.*

schachtii by calculation of temperature degree days. The rate of development is dependent upon soil temperature about 300 day degrees above a base of 10°C being required for completion of a generation (Greco et al. 1982). A generation cycle was completed after seven weeks and the trials were harvested. Fresh plant shoots were weighed (g). Cysts of *H. schachtii* were extracted from a soil subsample of 200 g according to Jenkins (1964) and counted. Furthermore, the number of eggs and juveniles within the cysts was recorded.

7.2.5 Effect of fungal Alginate ALG+PA5 capsules on sugar beet production and nematode reproduction of Heterodera schachtii

An experiment on sugar beet ›Dorena‹ with *H. schachtii* was conducted in 18 l pots in heat-treated field soil. The large volume of these pots enabled the sugar beet to grow a well-formed body. The pots were placed outside in a vegetation hall, where the plants were exposed to naturally fluctuating weather temperatures. The trial was set-up in the middle of April 2005. The 18 l pot trial was conducted with Alginate ALG+PA5 capsules containing *H. rhossiliensis* against *H. schachtii* on sugar beet. Product quality was certified in a vitality assay, which was conducted as previously described in chapter 7.2.3 »Efficacy of fungal Alginate ALG+PA5 capsules against *Meloidogyne incognita* on cucumber.

There were four treatments:
 (1) Non-treated control (CON)
 (2) Fungal biomass of *H. rhossiliensis* (H.r.)
 (3) Alginate ALG+PA5 capsules with 3% baker's yeast (ALG+PA5)
 (4) Alginate ALG+PA5 capsules with 3% baker's yeast and 1% *H. rhossiliensis* (ALG+PA5+H.r.)

Under natural conditions, eggs and juveniles remain in cysts and may hatch at temperatures between 16°C and 28°C also stimulated by the presence of radicular secretions of host plants. At trial set-up not only the treatment but also 489.8 g of cyst inoculum were added to 16 kg of heat-treated field soil to create a population density of 1500 eggs and juveniles per 100 g of soil. Each treatment had 11 replicates. The application rate of capsules per pot was 640 g of moist capsules per 16 kg of soil. As the capsules contained 1% *H. rhossiliensis*, the treatment consisted of 6.4 g fungal biomass. For the treatment »Fungal biomass of *H. rhossiliensis*« 6.4 g of biomass were suspended into 100 ml of water and mixed into the soil of each pot. Cysts were thoroughly mixed into the soil. To secure a homogenous distribution of capsules or fungal biomass and cysts of *H. schachtii* in the soil, the soil was mixed thoroughly in a cement mixer for 3 minutes before being filled into the 18 l pot. Irrigation was applied as needed. On day 14, six sugar beet pills ›Dorena‹ were sown approximately a centimetre deep in each pot. Two weeks after the seeds had germinated, all seedlings but one were removed from each pot, rinsed free of soil and the number of nematodes within a root system was assessed. After four months, the remaining sugar beet plant was harvested. Fresh upper plant weights and fresh weight of sugar beet roots was assessed. The number of *H. schachtii* eggs and juveniles in a subsample of 400 g was determined by extracting cysts from soil according to Jenkins (1964) and determining the amount of eggs and juveniles within the cysts per 100 g of soil.

7.3 Results

7.3.1 Influence of different soil types on growth of Hirsutella rhossiliensis

Results of the *in vitro* test conducted with an encapsulation of *H. rhossiliensis* in Alginate ALG+PA5 and fungal pellets consisting of mycelia of *H. rhossiliensis* showed that the factor heat-treated soil or non-treated soil had a much greater influence on growth of aerial mycelia than the different soil types. If Alginate ALG+PA5 capsules containing *H. rhossiliensis* were placed on WA, moist filter paper, heat-treated sand, heat-treated field soils with different humus contents or heat-treated loess, growth of aerial mycelia of *H. rhossiliensis* could be observed after merely two days. The same observations were made for fungal pellets. If Alginate capsules were placed on non-treated field soils A-C or non-treated loess, growth of aerial mycelia of *H. rhossiliensis* was inhibited. Sand was the only insterile substrate, on which fungal growth from Alginate ALG+PA5 capsules could be recorded. All other non-treated soil substrates contained microorganisms that infested the capsules and inhibited growth of *H. rhossiliensis*. On the contrary, growth of aerial mycelia of *H. rhossiliensis* from pure fungal pellets was not affected if placed on non-treated substrates. Aerial mycelia grew successfully from the fungal pellets, regardless of the substrate it was placed on and whether it was heat-treated or not. Figure 26 and Figure 27 display Alginate ALG+PA5 capsules containing *H. rhossiliensis* and fungal pellets of *H. rhossiliensis* eight days after trial set-up. Before ending the trial on day 15, the amount of aerial mycelia surrounding the capsule or pellet was assessed in form of ranking on a scale from 0-5 as described in chapter 7.2.1 »Influence of different soil types on growth of *Hirsutella rhossiliensis*«. Aerial mycelia did not grow from ALG+PA5 capsules containing *H. rhossiliensis* if placed on non-treated field soils A, B and C and non-treated loess, therefore growth was ranked 0. If ALG+PA5 capsules were placed on moist filter paper, non-treated sand and heat-treated field soil A, the median of growth was ranked either 3 or 4 meaning that about 60% or 80% of the capsule, respectively, visible to the viewer, was covered with aerial mycelia and 20% to 40% of the bare capsule was not covered with aerial mycelia. Fungal pellets on the other hand that had been placed on all substrates but non-treated field soil B and C were graded with a 5 (100%). Aerial mycelia grew densely from all of these pellets. As already stated above, fungal growth of aerial mycelia from pellets placed on non-treated field soil B and C was observed shortly after trial begin, yet after 12 days, the pellet began to vanish and on day 15, aerial mycelia was not visible and growth was rated 0 (Figure 28).

Heat-treated

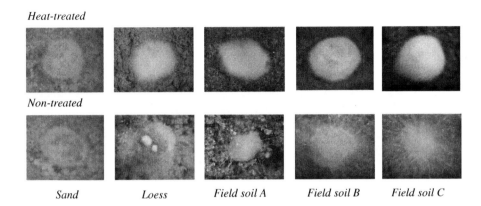

Non-treated

Sand   Loess   Field soil A   Field soil B   Field soil C

Figure 26: **Alginate ALG+PA5 capsules** *containing* Hirsutella rhossiliensis *with aerial mycelium (top pictures) or infested with antagonistic fungi (bottom pictures) eight days after placement on the following substrates: (top pictures display heat-treated soil substrates, bottom pictures display non-treated soil substrates) from left to right: Sand, loess, field soil A, field soil B and field soil C.*

Heat-treated

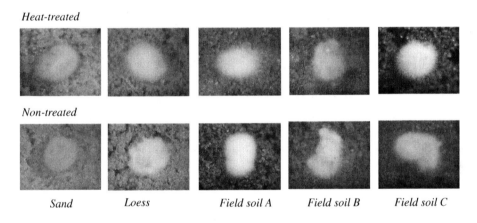

Non-treated

Sand   Loess   Field soil A   Field soil B   Field soil C

Figure 27: **Fungal pellets** *of* Hirsutella rhossiliensis *with aerial mycelium (top pictures) eight days after placement on the following substrates: (top pictures display heat-treated soil substrates, bottom pictures display non-treated soil substrates) from left to right: sand, loess, field soil A, field soil B and field soil C.*

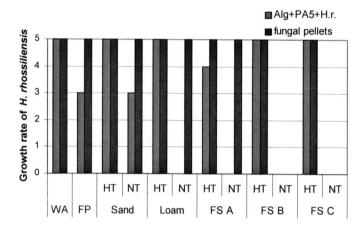

Figure 28: The growth rate of aerial mycelia of Hirsutella rhossiliensis *displayed as the median from Alginate ALG+PA5 capsules containing* H. rhossiliensis *and fungal pellets of* H. rhossiliensis *placed on 12 different substrates on a scale of 0 to 5 (0 = 0% visible growth, 1 = growth of aerial mycelia from 20% of the capsule or pellet surface, visible to the viewer, 2 = 40%, 3 = 60%, 4 = 80%, 5 = 100%), WA = water agar (1.5%), FP = moist filter paper, HT = heat-treated, NT = non-treated, FS = field soil; n = 30.*

7.3.2 Efficacy of fungal Alginate ALG+PA5 capsules against Heterodera schachtii on sugar beet

Vitality assays showed that Alginate ALG+PA5 capsules with and without *H. rhossiliensis* were contaminated by other microorganisms. As the vitality test to confirm product quality was conducted on the same day as the 100 ml biotest trial set-up, the trial was still carried out and evaluated. The influence of capsule components ALG and PA5, respectively, could still be investigated. Results of an application of contaminated ALG+PA5 capsules with and without *H. rhossiliensis* were displayed in the graphic but were not taken into consideration when the results of the trial were described (Figure 29).

The capsule component Alginate ALG as a powder did not significantly affect root penetration with *H. schachtii* juveniles in comparison to the non-treated control, whereas, the application of the pectine derivate PA5 in form of a powder had a negative influence on nematode invasion of *H. schachtii* into the roots. Nematode invasion increased by over 100% in comparison to the non-treated control. In plants treated with *H. rhossiliensis* as a liquid suspension, root invasion was 13% lower than in non-treated plants, but the differences were not significant (Figure 29).

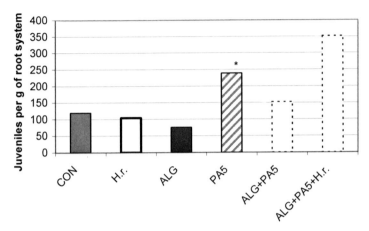

Figure 29: Effect of Alginate ALG+PA5 capsules and capsule components in heat-treated field soil on nematode invasion of Heterodera schachtii *into the roots of sugar beet seedlings. CON = Non-treated control, H.r. = Biomass of* Hirsutella rhossiliensis *as a liquid suspension, ALG = Alginate powder, PA5 = Pectine derivate powder, ALG+PA5 = Alginate ALG+PA5 capsules (contaminated), ALG+PA5+H.r. = Alginate ALG+PA5 capsules containing* H. rhossiliensis *(contaminated). Values are means of 12 replicate pots, * = significantly different to control according to Tukey HSD-Test with p ≤ 0.05.*

7.3.3 Efficacy of fungal Alginate ALG+PA5 capsules against Meloidogyne incognita on cucumber

Results of the vitality test confirmed product quality of Alginate ALG+PA5 capsules containing *H. rhossiliensis*. Under sterile conditions aerial mycelia of *H. rhossiliensis* grew successfully from the capsules within two days.

Although product quality was given, Alginate ALG+PA5 capsules containing *H. rhossiliensis* did not significantly affect root penetration by second-stage juveniles of *M. incognita* whether applied to heat-treated or non-treated soil in comparison to the non-treated control (Figure 30 and Figure 31). In heat-treated soil, a capsule application even increased nematode invasion by 38%, although this difference was not statistically significant (Figure 30). In non-treated soil, an application of Alginate ALG+PA5 capsules containing *H. rhossiliensis* reduced root penetration by 46%. The difference was not statistically significant in comparison to the non-treated control (Figure 31). An application of *H. rhossiliensis* as a liquid suspension did not lead to a significant reduction of nematode invasion into the roots of cucumber seedlings, whether applied to heat-treated or non-treated soil (Figure 30 and Figure 31).

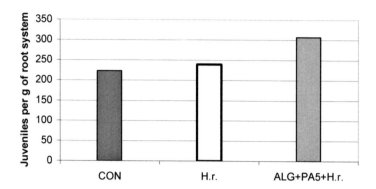

Figure 30: Effect of Alginate ALG+PA5 capsules containing 5% Hirsutella rhossiliensis *(ALG+PA5+H.r.) applied in heat-treated soil on nematode invasion of* Meloidogyne incognita *in the roots of cucumber seedlings. CON = Non-treated control, H.r. = Biomass of* H. rhossiliensis *as a liquid suspension. Values are means of 12 replicate pots, no significant differences according to Tukey HSD-test with p ≤ 0.05.*

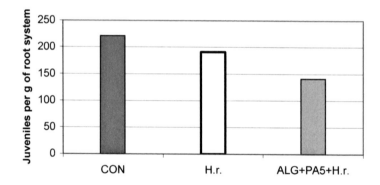

Figure 31: Effect of Alginate ALG+PA5 capsules containing 5% Hirsutella rhossiliensis *(ALG+PA5+H.r.) applied in non-treated soil on nematode invasion of* Meloidogyne incognita *in the roots of cucumber seedlings. CON = Non-treated control, H.r. = Biomass of* H. rhossiliensis *as a liquid suspension. Values are means of 12 replicate pots, no significant differences according to Tukey HSD-Test with p ≤ 0.05.*

7.3.4 Effect of fungal Alginate ALG+PA5 capsules on nematode reproduction of Heterodera schachtii

Alginate ALG+PA5 capsules containing the endoparasitic fungus *H. rhossiliensis* were not contaminated by other microorganisms prior to trial set-up. The fungus grew from 9 out of 10 capsules successfully within 3 days.

In heat-treated soil, a capsule application of Alginate ALG+PA5 capsules with and without *H. rhossiliensis* reduced plant shoot fresh weight by over 32% when compared to the non-treated control (Figure 32). An application of *H. rhossiliensis* as a fungal suspension increased upper plant shoot fresh weight by over 20% in comparison to non-treated plants. The differences however were not statistically significant.

In heat-treated field soil, an application of Alginate ALG+PA5 capsules with and even without *H. rhossiliensis* significantly reduced the amount of cysts of *H. schachtii* by over 37% and the amount of eggs and juveniles produced by over 41% per 100 g of soil in comparison to the non-treated control. *H. rhossiliensis* as a liquid suspension, significantly decreased the amount of cysts per 100 g of soil by over 40%, however, the number of eggs and juveniles was increased by 17% in comparison to the non-treated control, although this difference was not statistically significant (Figure 32).

In non-treated soil, all three treatments did not affect fresh plant shoot fresh weight of sugar beets when compared to the control (Figure 32). Also, an application of *H. rhossiliensis* as a liquid suspension did not affect the amount of cysts per 100 g of soil nor did it affect the amount of eggs and juveniles produced in comparison to the non-treated control. An application of Alginate ALG+PA5 capsules with and without *H. rhossiliensis* led to a reduction of cysts of over 25%, although the difference was not significant. Even though the amount of cysts was reduced, the amount of eggs and juveniles within the cysts was not. A treatment of capsules containing *H. rhossiliensis* led to an increase of eggs and juveniles by 50% per 100 g of soil in comparison to the control. This difference was not statistically significant (Figure 32).

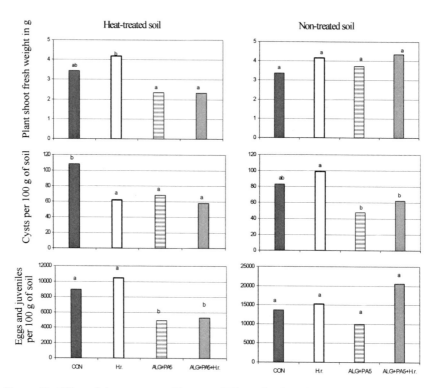

Figure 32: Effect of the treatments biomass of Hirsutella rhossiliensis *as a liquid suspension (H.r), Alginate ALG+PA5 capsules (ALG+PA5), Alginate ALG+PA5 capsules containing 5% H.* rhossiliensis *(ALG+PA5+H.r.) and a non-treated control (CON) on plant shoot fresh weight of sugar beet ›Dorena‹, on the number of cysts of* Heterodera schachtii *and the number of eggs and juveniles of* H. schachtii *per 100 g of soil, either heat-treated soil (graphs on the left hand-side) or non-treated soil (graphs on the right hand-side). Values are means of 12 replicate pots; Means followed by the same letter are not significantly different according to Tukey HSD-Test with p ≤ 0.05.*

7.3.5 Effect of fungal Alginate ALG+PA5 capsules on sugar beet production and nematode reproduction of Heterodera schachtii

Vitality assays showed that delivered Alginate ALG+PA5 capsules with 1% *H. rhossiliensis* did not have contaminations. Radial growth of aerial mycelia of *H. rhossiliensis* out of capsules was detected on WA, filter paper and heat-treated soil.

Results of the 18 l pot trial conducted in heat-treated soil showed that an application of moist Alginate ALG+PA5-capsules containing 1% fungal biomass led to a significant reduction of nematode invasion of 58% into the roots of two-week old sugar beet seedlings in comparison to the non-treated control (Figure 33). In contrary, an application of Alginate ALG+PA5 capsules without fungus significantly increased root penetration by juveniles of *H. schachtii* by over 70% when compared to the control. Nematode invasion in plant roots treated with *H. rhossiliensis* as a liquid suspension was 64% lower than the non-treated plant roots, which was comparable to an application of Alginate ALG+PA5-capsules containing 1% *H. rhossiliensis* (Figure 33).

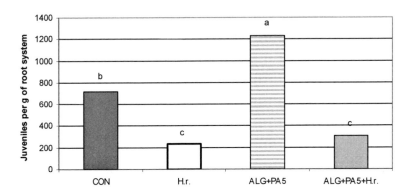

Figure 33: Effect of the treatments Hirsutella rhossiliensis as a liquid suspension (H.r), Alginate ALG+PA5 capsules (ALG+PA5) and Alginate ALG+PA5 capsules containing 1% H. rhossiliensis (ALG+PA5+H.r.) applied in heat-treated soil and a non-treated control (CON) on nematode invasion of Heterodera schachtii into the roots of sugar beet seedlings two weeks after germination. Values are means of 12 replicate pots. Means followed by the same letter are not significantly different according to Tukey HSD-Test with p ≤ 0.05.

At harvest, there were distinct differences in size and morphology of sugar beet roots among the different treatments. Non-treated roots of sugar beet plants weighed significantly less than the roots of treated plants and had severe nematode damage in form of root branching (90% of all roots) (Figure 34 and Figure 35). In contrast, a treatment with *H. rhossiliensis* as a liquid suspension led to well-formed roots with little or no root branching. An application of Alginate ALG+PA5-capsules without *H. rhossiliensis* led to larger roots but roots were bearded and branched. The highest root weight with an average of 199 g and the least amount of root branching (30%) was achieved by an application of Alginate ALG+PA5 capsules containing 1% *H. rhossiliensis* (Figure 34 and Figure 35).

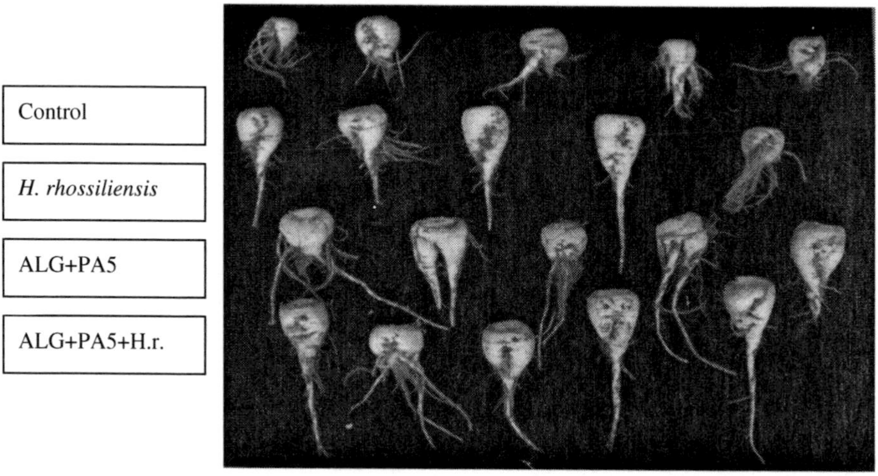

Control

H. rhossiliensis

ALG+PA5

ALG+PA5+H.r.

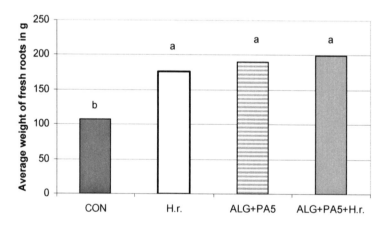

Figure 34: Influence of different treatments with and without Hirsutella rhossiliensis *on the development of sugar beet roots. Five characteristical sugar beet roots of each treatment are presented. From top to bottom: Non-treated control (Control);* H. rhossiliensis *as a liquid suspension (*H. rhossiliensis*); ALG+PA5 = Alginate ALG+PA5-capsules; ALG+PA5+H.r. = Alginate ALG+PA5-capsules containing 1%* H. rhossiliensis

Figure 35: Effect of the treatments biomass of Hirsutella rhossiliensis *as a liquid suspension (H.r), Alginate ALG+PA5 capsules (ALG+PA5) and Alginate ALG+PA5 capsules containing 1%* H. rhossiliensis *(ALG+PA5+H.r.) applied in heat-treated soil and a non-treated control (CON) on the average weight of fresh roots in g at harvest. Values are means of 11 replicate pots; Means followed by the same letter are not significantly different according to Tukey HSD-Test with* $p \leq 0.05$.

Nematode reproduction did not decrease, although an application of Alginate ALG+PA5 capsules containing *H. rhossiliensis* reduced root penetration into two-week old sugar beet seedlings and increased the average weight of sugar beet roots at harvest. Soil samples taken at harvest showed that all of the treatments did not lead to a significant reduction in the reproduction of eggs and juveniles in comparison to the control. A treatment of ALG+PA5 capsules with and without *H. rhossiliensis* even led to a significant increase of eggs and juveniles of *H. schachtii*. An application of Alginate ALG+PA5 capsules without *H. rhossiliensis* significantly increased nematode reproduction by 1035%. An application of capsules containing the fungus led to a significant increase of 386% when compared to the non-treated control (CON). A treatment of *H. rhossiliensis* as a liquid suspension increased the number of eggs and juveniles by 54%, although the increase was not statistically significant in comparison to the control (Figure 36).

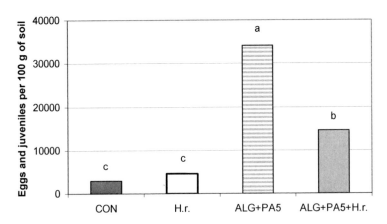

Figure 36: Effect of the treatments Hirsutella rhossiliensis as a liquid suspension (H.r), Alginate ALG+PA5 capsules (ALG+PA5) and Alginate ALG+PA5 capsules containing 5% H. rhossiliensis (ALG+PA5+H.r.) applied in heat-treated soil and a non-treated control (CON) on nematode reproduction at harvest. Values are means of 11 replicate pots; Means followed by the same letter are not significantly different according to Tukey HSD-Test with p ≤ 0.05.

7.4 Discussion

In the *in vitro* test conducted with five different types of soil, the formulation of the fungus had the greatest affect on growth of aerial mycelia of *H. rhossiliensis*. Alginate ALG+PA5 capsules with 5% *H. rhossiliensis* exhibited radial growth of the fungus only if capsules were placed on heat-treated or sterile substrates and non-treated sand. In contrary, growth of aerial mycelia from unformulated pure fungal pellets was registered, regardless if the soil substrate had been heat-treated or not. The abiotic factor soil type did not affect fungal growth of *H. rhossiliensis* from the capsules to the extent the biotic factors fungal and bacterial microorganisms did. Also, Jaffee et al. (1996) did not believe that abiotic conditions of a variety of soils was responsible for the lack of control by Alginate capsules containing *H. rhossilien-*

sis placed in field soil but that biotic factors played a more important role. Capsules placed on non-treated soils were infested by microorganisms as shown in Figure 26, which prevented growth of *H. rhossiliensis*. No fungicides or bactericides were added to the formulation of Alginate ALG+PA5 capsules containing *H. rhossiliensis* to limit contamination. Fungal pellets on the other hand were not infested by other microorganisms and sporulation occurred from the fungal pellet which consisted of assimilative hyphae that were grown in shake culture and rinsed free of media. Sporulation involved converting previously assimilated food into new hyphae, phialides, and conidia (Jaffee, 2000). Lackey et al. (1992) reported that vegetative colonies sporulated and infected J2 of *H. schachtii* equally well in non-heated, 60°C heated and autoclaved soils. In our study, after 12 days, the pellets, that had been placed on non-treated field soils B and C started to diminish. A reason for this might be that the assimilative resources of the pellet had been depleted. Perhaps, the fungal pellet could be compared to a parasitized *H. schachtii* juvenile. Sporulation from a juvenile requires fourteen days and after fourteen days the nematode substrate has been depleted and no new spores are produced (Jaffee et al., 1990).

Results of the 100 ml biotest conducted on sugar beet against *H schachtii* with contaminated Alginate ALG+PA5 capsules with and without 5% *H. rhossiliensis* in heat-treated field soil showed that the capsule component alginate (ALG) did not have an effect on root penetration, whereas pectine derivate (PA5) as a powder did have a stimulating affect on root penetration of *H. schachtii* juveniles. As Robinson and Jaffee (1996) stated that alginate capsules containing *Monacrosporium conopagum*, *M. ellipsosporum* and *H. rhossiliensis* repelled J2 of *Meloidogyne incognita* in a behavioural assay, it was hypothesized that the capsule component alginate (ALG) in form of a powder might have an affect on root penetration in our study but nematode invasion did not differ significantly to the control. If further trials verify the stimulating effect pectine derivate PA5 has on nematode invasion into roots, it should be counteracted by an attractant added to the formulation that manipulates nematode behaviour and lures the nematode to the capsule and not away from the capsule towards the root system.

In the 100 ml biotest conducted on sugar beet against *H. schachtii* in heat-treated field soil and also in the 100 ml biotest conducted on cucumber against *Meloidogyne incognita* in heat-treated and non-treated field soil, a liquid suspension of *H. rhossiliensis* was not effective for control of the plant-parasitic nematodes *H. schachtii* or *M. incognita*, respectively. It is difficult to explain why *H. rhossiliensis* sometimes achieves a good efficacy as a liquid suspension and why in other studies there is no efficacy at all. Furthermore, an application of Alginate ALG+PA5 capsules containing *H. rhossiliensis* did not significantly reduce nematode invasion of *M. incognita* into the roots of cucumber seedlings in heat-treated field soil. One of many possible explanations for the lack of efficacy could be that the soil was too moist for aerial mycelia of *H. rhossiliensis* to grow. One-hundred ml trials are conducted in plastic rectangular containers that do not allow for easy drainage of excess water. Viaene and Abawi (2000) also observed that frequent high moisture content of soils in the pots with *H. rhossiliensis*, probably had a negative effect on spore transmission and sporulation of *H. rhossiliensis*. Tedford et al. (1992) found that soil water status substantially affected transmission of *H. rhossiliensis* to *H. schachtii* and *M. javanica* in loess, loessy sand and sand. Furthermore, they found that water levels that affected transmission did not affect motility of *H. schachtii* in loessy sand, according to the results of a root-penetration assay, which could explain why root penetration *H. schachtii* and *M. incognita* in our studies was not affected, regardless of the treatment. In non-treated soil, abiotic factors fungal and microorganisms additionally effect fungal growth from capsules.

In the 300 ml pot trial conducted with *H. schachtii* on sugar beet in heat-treated soil, there were no differences in the efficacy of Alginate ALG+PA5 capsules with and without *H. rhossiliensis*. Although both treatments significantly reduced the amount of cysts produced including the amount of eggs and juveniles per 100 g of soil, the efficacy certainly cannot be based on parasitism of *H. schachtii*. Although plant growth in heat-treated soil was not significantly reduced by an application of Alginate ALG+PA5 capsules with and without *H. rhossiliensis*, it did cause a reduction of plant growth of over 32%. Perhaps the high amount of Alginate capsules (12 g) applied to 300 cc of heat-treated soil influenced plant growth negatively and consequently also nematode development as observed by Schuster and Sikora (1992). Furthermore, decomposition of the capsules may also have affected nematodes. As plant growth was not inhibited by *H. rhossiliensis* as a liquid suspension, the significant reduction of cysts indicates that an application of fungal suspension suppressed penetration of sugar beet roots by *H. schachtii* and therefore reduced the amount of cysts produced. Due to the fact that plant growth was not inhibited, the juveniles that entered the root system were provided with enough nutrients to develop into well nourished females resulting in a high number of produced eggs. This might explain why the number of eggs and juveniles was not significantly reduced when compared to the non-treated control. Alginate capsules with and without *H. rhossiliensis*, as well as a liquid treatment of *H. rhossiliensis*, that were applied to non-treated field soil did not influence plant growth, nor did the treatments reduce the amount of produced cysts and the containing amount of eggs and juveniles per 100 g of soil. Although biotic inhibition could be the cause for failure of the alginate capsule (ALG+PA5+H.r.) formulation containing *H. rhossiliensis*, it is still questionable why the performance of *H. rhossiliensis* as a liquid suspension was so poor in this study. Liu and Chen (2005) demonstrated that liquid cultures of *H. rhossiliensis* and *H. minnesotensis* were highly effective in control of the soybean cyst nematode *Heterodera glycines* when applied at 0.2-0.8 g of mycelium/300 ml pot. The 300 ml pots in our study were treated with 0.6 g of mycelium/300 ml pot which should have been an effective amount for control of the sugar beet cyst nematode. Once again, the fungus is sensitive to high soil moisture (Tedford et al. 1992) which limits its growth. Furthermore, Lackey et al. (1992) explained that temporal delay associated with sporulation from vegetative colonies may cause substantial numbers of endoparasitic nematodes to infect roots after fungal colonies were added to the soil but before high densities of spores were present. Even when spores are present, endoparasitic nematodes that hatch and travel only short distances to host roots may escape parasitism.

91

Perhaps these factors contributed to the ineffectiveness of *H. rhossiliensis* as a liquid suspension.

The 18 l pot trial conducted with sugar beet against *H. schachtii* in heat-treated soil demonstrated that an application of *H. rhossiliensis* as a liquid suspension was highly effective in control of the nematode. As sugar beet seeds were sown two weeks after soil treatment with the fungal suspension and an additional 4 days were needed for the seedlings to germinate, *H. rhossiliensis* had sufficient time to sporulate in the soil. The 18 l pots had been placed in a vegetation hall under a protective roof, so that the soil moisture could be controlled. Furthermore, 18 l pots did not need to be irrigated as often as 100 ml boxes and 300 ml pots and were better drained, thus promoting growth of *H. rhossiliensis* perhaps. In California peach orchards, *H. rhossiliensis* parasitized more nematodes in drier soil between irrigation furrows than in wetter soil adjacent to irrigation furrows (Jaffee et al. 1989). Nematode invasion was reduced by 58% in the roots of two-week old sugar beet seedlings by a treatment of *H. rhossiliensis* as a fungal suspension. Consequently, a root system that is protected in the early stages of development can grow into a bigger and less damaged root system in comparison to a root system that was not protected. Therefore, the non-treated roots or sugar beet bodies showed much more severe signs of nematode damage in comparison to roots treated with a liquid suspension of *H. rhossiliensis*. A treatment led to larger roots and less nematode damage by 70%. An application of Alginate capsules containing *H. rhossiliensis* also led to a great reduction of nematode penetration in comparison to the non-treated control, whereas capsules without the fungus stimulated and enhanced root penetration of *H. schachtii*. This shows that the environmental conditions of this study were different to the studies described previously, in which alginate capsules with and without *H. rhossiliensis* had the same efficacy. Roots of harvested sugar beets treated with capsules that did not contain *H. rhossiliensis*, were large but severely damaged. The application of empty capsules containing baker's yeast increased plant growth but did not prevent nematode damage, whereas a treatment of capsules containing *H. rhossiliensis* also promoted plant growth but in addition improved plant health. Nematode damage was also reduced by 70% in comparison to the control. Although plant health was improved and sugar beets were protected to a certain extent, a reduction in the amount of eggs and juveniles per 100 g of soil was not achieved at harvest. Even though nematode invasion was reduced in the early stages of development, the amount of eggs and juveniles reproduced were not. Not even the fungal suspension reduced the amount of eggs and juveniles reproduced at harvest. Researchers have reported successes in suppressing nematode invasion of plant-parasitic nematodes into the roots of seedlings by a treatment of *H. rhossiliensis* as in our 18 l pot trial (Velvis and Kamp, 1996; Viaene and Abawi, 2000) and Liu and Chen (2005) demonstrated that liquid suspensions of *H. rhossiliensis* were able to reduce the percentage of eggs/cm3. Unfortunately, such results were not obtained in this trial.

The hypothesis that conducting trials in heat-treated soil would lead to a consistency in performance of the Alginate ALG+PA5 capsules by limiting the biotic factor microbial inhibition was not verified in these studies. Furthermore, the high inoculation level (4 g/100 g of soil), which is not practical for commercial use in a field or a greenhouse, was not effective enough for a biocontrol use.

7.5 Summary

The efficacy of Alginate ALG+PA5 as encapsulation material for *Hirsutella rhossiliensis* was investigated in a series of studies. An *in vitro* test was conducted with Alginate ALG+PA5 capsules and pure fungal pellets which were placed on five different soil types that were either heat-treated or non-treated. Results showed that fungal and bacterial microorganisms in non-treated soil types suppressed growth of aerial mycelia of *H. rhossiliensis* from ALG+PA5 capsules but did not effect growth from fungal pellets. Results of greenhouse pot trials (100 ml biotest, 300 ml pot trial) with *Heterodera schachtii* and *Meloidogyne* spp. conducted in non-treated soil also revealed that growth of *H. rhossiliensis* was inhibited by bacterial and fungal antagonists, thus a treatment with fungal capsules failed to suppress the impact of nematodes. Results of trials conducted in heat-treated soil did not exhibit contamination of capsules by other microorganisms, however, growth of *H. rhossiliensis* was suppressed by abiotic factors presumably. Again, nematode numbers were not decreased. Best results were obtained in the 18 l pot trial conducted with sugar beet against *H. schachtii* in heat-treated soil. The impact of *H. schachtii* on sugar beet roots was reduced in comparison to the non-treated control. Sugar beet roots were not only larger but also showed less signs of nematode damage. Unfortunately, a reduction of nematode reproduction was not achieved at harvest. As an application of ALG+PA5 capsules containing *H. rhossiliensis* should primarily reduce the population density of plant-parasitic nematodes in field soil, results achieved in these studies were not satisfactory. Capsules are very susceptible to contamination and *H. rhossiliensis* is a weak competitive saprophyte that reacts sensitively to environmental factors which have an impact on its growth. Thus, an encapsulation of *H. rhossiliensis* as a formulation for the fungus shows many negative side-effects that limit the capsule's efficacy. A successful application of *H. rhossiliensis* as a method of biological control requires the development of an optimised encapsulation or perhaps even an alternative formulation.

8 Effect of additives to enhance the biocontrol potential of encapsulated *Hirsutella rhossiliensis* towards *Heterodera schachtii* and *Meloidogyne incognita*

8.1 Introduction

To enhance the biocontrol efficacy of encapsulated Hirsutella rhossiliensis against infective juveniles of Heterodera schachtii and Meloidogyne incognita, attractants can be added to the capsule to lure infective juveniles to the capsule with sporulating H. rhossiliensis and thereby increase the infection rate of juveniles. The concept »attraction« is frequently employed in connection with orientation purposes (Klinger, J. 1965). Dethier (1947) defines the concepts of »attractant« and »repellent« as follows: Any stimulus which elicits a positive directive response may be termed as attractant; any stimulus which elicits an avoiding reaction may be termed repellent. Attraction of plant-parasitic nematodes by plant roots may be the key to host recognition, specificity, and subsequent infection (Dusenbery, 1987; Steiner, 1925). Castro et al. (1990) suggested that environmentally tolerable repellents offered an alternative to pesticides in the protection of plants from plant-parasitic nematodes. Results of their research implied that salts beneficial as plant nutrition, if suitably applied, could be used to shield roots from nematode attack. This was suggested as a new means of plant protection. We hypothesized that though plants would be protected from nematode infestation, nematodes would not actively be controlled. Our research focused on the protection of plants by parasitizing plant-parasitic nematodes with the biocontrol organism H. rhossiliensis. This fungus is embedded in a capsule which is encorporated in the soil. The capsule is supposed to act as a protective shield for the fungus and supply the fungus with nutrients for an optimal growth. Once the fungus produces aerial mycelia and sporulates, nematodes must pass by the sticky spores of H. rhossiliensis for an attachment to occur to the cuticle of the nematode. Only then can infection and parasitism begin. The addition of an attractant to the capsule formulation could lure nematodes away from the plant and lead them straight to the endoparasitic fungus to secure a high rate of infection. Theoretically, if an optimal attractant is found, the application rate of capsules containing H. rhossiliensis could be drastically reduced as nematodes would wander to the capsule with sporulating H. rhossiliensis, become infested and parasitized and act as new sites for infestation. Additives were chosen that could potentially have attractant activity. In linked twin-pot chamber trials, consisting of two connecting pots, the aggregation of infective juveniles as a result from directed migration into the treated pot (the source of stimulation) or treated root was investigated in comparison to the non-treated control.

8.2 Material and Methods

8.2.1 Experimental design of Linked Twin-pot Chamber trials

The sensitivity of the nematode species *H. schachtii* or *M. incognita*, towards various additives was investigated in Linked Twin-pot Chamber trials (Figure 37). A selected additive was tested in two different trials. Each trial consisted of five sets of Linked twin-pot Chambers. The 8 cm Twin-pots were filled with 200 ml of silicate sand. The two pots were connected by a tube, which was also filled with sand. The tube was constructed out of plastic (1.3 x 1.5 x 5 cm). The hole at the middle of the tube, which served as the spot for inoculation of second-stage juveniles of *H. schachtii* or *M. incognita*, was 5 mm in diameter. At day 0, the right pot was treated with an additive that could potentially be an attractant of the sugar beet cyst nematode or the root-knot nematode, respectively. The sand of the left pot was treated with water. In the first experiment for Linked Twin-pot Chamber, pots remained without seeds. Five thousand second-stage juveniles of *H. schachtii* or *M. incognita* in approximately 1 ml of water were inoculated through a hole at the middle of the tube on day 0. Juvenile inoculum was obtained as described in chapter 2.2.2. and 2.2.3 »Preparation of juvenile inoculum of *Heterodera schachtii/Meloidogyne incognita*«. At day six, juveniles of *H. schachtii* or *M. incognita* were extracted from the pots and the connecting tube and were recorded. In a second experiment, two sugar beet seeds ›Dorena‹ (for *H. schachtii*) or tomato seeds ›UC82 Davis‹ (for *M. incognita*) were sown per pot of the Linked Twin-pot Chamber trials conducted for each additive. At day ten, nematode infestation into the roots of sugar beet seedlings or tomato seedlings of all plants of the second Linked Twin-pot Chamber trial was determined by root staining (Byrd et al. 1983). Experiments were placed in the greenhouse at 20°C ± 3°C.

5000 J2

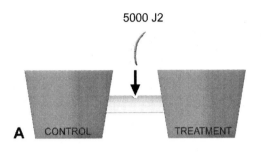

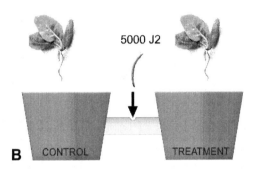

Figure 37: Experimental design of the Linked Twin-pot Chamber used to determine the attractant activity of various additives on Heterodera schachtii *or* Meloidogyne incognita *attraction, respectively (A) and the influence of these additives on penetration into roots of sugar beet ›Dorena‹ seedlings or tomato ›UC 82 Davis‹ seedlings, respectively (B).*

8.2.2 Sensitivity of Heterodera schachtii *towards potential additives of capsules*

The following additives that could potentially have attractant activity on second-stage juveniles of *Heterodera schachtii* were chosen for the Linked Twin-pot Chamber trials.

Hirsutella rhossiliensis

To determine whether the active ingredient of the capsule had any effect on nematode movement, a fungal suspension of *H. rhossiliensis* (0.4 g of vegetative hyphae/4 ml of water/per pot) was inoculated on the surface of the sand of the right pot, whereas the left pot was treated with 4 ml of water and used as controls.

Macerated sugar beet ›Monza‹

One gram of the sugar beet variety ›Monza‹ (Hilleshög, Syngenta, Basel, Switzerland), which had been macerated in a blender for a few seconds prior to trial set-up, was filled into a pocket made of gauze (3.5 cm x 3.1 cm; mesh = 100 µm). The opening of the pocket was sealed off and the gauze was placed in the centre of each pot and covered with sand. A pocket made of gauze without macerated sugar beet was placed in the control pots. At trial take down, in addition to the standard parameter described previously, the gauze pockets containing macerated sugar beet were placed on Baermann funnels to collect juveniles of *H. schachtii* that might have entered the pockets.

Rape root exudates

Root exudate from rapeseed (*Brassica napus*) promotes hatching of cysts of *H. schachtii*. Thus, the influence of root exudates on nematode movement was investigated in these trials. The roots of approximately 250 five-to-six-day-old rapeseed ›Jimbo‹ seedlings were placed in a glass beaker filled with 200 ml of distilled water for 24 hours. The plants were removed and the solution consisting of water with root exudates of rape seedlings was sterile filtered. Two trials were set-up parallel for each Linked Twin-pot chamber trial. In the first trial, right pots were treated with 1 ml of root exudates solution at day 0, whereas left pots received 1 ml of water and were used as controls. In the second trial, right pots were treated two times with 1 ml of root exudate solution (on day 0 and day 3) whereas left pots received 1 ml of water at each application.

Trehalose

The nutritional requirement of *H. rhossiliensis* was investigated by Chen et al. (2004). Carbohydrates and nitrogen compounds are needed for growth, sporulation and spore germination of *H. rhossiliensis*. D (+) trehalose was found to be a very good carbohydrate and nitrogen source for sporulation of *H. rhossiliensis*. For this reason trehalose was chosen as a possible additive for the capsule containing *H. rhossiliensis*. Trehalose ($C_{12}H_{22}O_{11}$ x 2 H_2O), also known as mycose, is an alpha-linked sugar found extensively but not abundantly in nature. It can be synthesised by fungi, plants and invertebrate animals. Trehalose was applied at 150 mg per pot of sand. The control was left non-treated.

Vanillic Acid

Vanillic acid ($C_8H_8O_4$ – Merck, Hohenbrunn, Germany) is a component of natural vanilla extract. Jaffe et al. (1989) identified vanillic acid as an attractant for males of the cyst nematode *Heterodera glycines*. In the Linked Twin-pot Chamber trials, vanillic acid was tested as a potential attractant of *H. schachtii*. A rate of 0.3 mg of vanillic acid was applied to 200 g of sand, thoroughly mixed and filled into the right pot of the Linked Twin-pot Chambers. The left pot remained non-treated and was used as a control.

Propatox 2%®

Propatox 2%® (1,1-Diacetoxy-2-ethylhex-2-en) is a chemical product, which is used to induce egg hatching of the plant-parasitic nematode genera *Heterodera* spp. for experimental purposes. This product was patented by Dr. Banasiak and Dr. Große of the Federal Biological Research Centre for Agriculture and Forestry (BBA), Kleinmachnow, Germany. Trials were conducted to investigate whether this product had an effect on nematode behaviour or not. A cosmetic pad was inoculated with 200 μl of Propatox 2%® and placed at the bottom of the right pot. Sand was then filled into the pot. For the control pot, 200 μl of water was applied to the cosmetic pad, which was placed at the bottom of the control pot and then covered with 200 g of sand. As Propatox 2%® becomes volatile when it comes into contact with water, it was necessary for the pots to remain moist at all times. A plastic hood was placed over the pots to prevent the water from evaporating too quickly.

8.2.3 Selected additive formulation for trials with Heterodera schachtii and Meloidogyne incognita

»Baker's yeast with OZ1«

The additive »OZ1« was developed by Dr. Anant Patel of the Federal Research Centre for Agriculture (FAL), Braunschweig, Germany. The substance »OZ1« was added to baker's yeast to create a formulation. The effect of this formulation on nematode movement was studied. Furthermore, the effect of an »empty capsule formulation« without the additive formulation and »capsules with baker's yeast and OZ1« on nematode movement was investigated to compare the influence of the testing materials and their efficacy in combination. The treatments were as follows:

(1) »Empty capsule formulation«
(2) »Baker's yeast with OZ1«
(3) »Capsules with baker's yeast and OZ1«.

The trials were conducted with *Heterodera schachtii* on sugar beet or on its own and with *Meloidogyne incognita* on tomato ›UC 82 Davis‹ or on its own. Trials were carried out as described in chapter 8.2.1 Experimental design of »Linked Twin-pot Chamber trials«. Right pots were treated with 4 g of an »empty capsule formulation« or 4 g of »Capsules with baker's yeast and OZ1«, whereas left pots remained non-treated. For the treatment »Baker's yeast with the OZ1«, 1 g of the substance was dissolved in 3 ml of water and applied to the right pot. The left pot was used as a control.

8.3 Results of Linked Twin-pot Chamber trials

8.3.1 Sensitivity of Heterodera schachtii *towards potential additives of capsules*

Of the total amount of nematodes extracted from the first Linked Twin-pot Chamber trials (both pots and tube), a relative amount of nematodes was found in the sand of the connecting tube. This percentage indicates the number of nematodes that did not migrate into either a treated or non-treated pot. The percentage varies depending on the selected additive applied to a trial. The relative amount of nematodes that did not migrate is listed in table 5.

Table 5: Percentage (%) of juveniles of Heterodera schachtii *that remained in the tube and did not migrate into either a treated or non-treated pot of the first Linked Twin-pot chamber trials conducted with various additives.*

Additive	(%) Juveniles in tube
H. rhossiliensis	55
Rapeseed exudates	30
Macerated sugar beet	40
Trehalose	63
Vanillic acid	36
Propatox 2%®	15
Control	27

The influence of selected additives on host attraction towards *H. schachtii* was studied in Linked Twin-pot Chamber trials without the presence of a host plant. Results indicated, that six days after inoculation of juveniles of *H. schachtii*, nematode attraction to treated pots was not promoted by an application with any selected additive. The additives had either no effect on nematode migration or a repellent one. An application of 1 g of macerated sugar beet significantly repelled juveniles of *H. schachtii* by 78% and an application of 0.3 mg of vanillic acid per 200 g of sand repelled nematodes by 44% (Figure 38). An application of *H. rhossiliensis* as a fungal suspension, root exudates of rape seedlings, Trehalose in form of a powder or Propatox 2%® did not have an influence on nematode migration from the tube into either the treated or non-treated pot. There was no indication of any sort of stimulation on nematodes from each of the four different additives. In the non-treated Linked Twin-pot Chamber that served as controls, no difference between nematode numbers of each pot was detected (Figure 38).

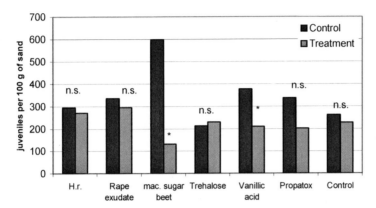

Figure 38: Effect of Hirsutella rhossiliensis *(H.r.), root exudates of rape seedlings (Rape exu-date), macerated sugar beet (mac. sugar beet), Trehalose, Vanillic acid, Propatox 2%® and Non-treated Control on the* **total number of juveniles of** Heterodera schachtii **in treated or non-treated** *pots in first Linked Twin-pot Chamber trial (per 100 g of sand). Means with (*) are sig-nificantly different based on T-Test (P ≤ 0.05; n = 5).*

Linked Twin-pot-Chamber trials conducted with sugar beet seedlings ›Dorena‹ showed that the presence of a seedling could additionally influence the activity of a nematode and its movement into a treated or non-treated pot. Furthermore, a treat-ment could also have phytotoxic effects on sugar beet seedlings. Trials conducted with seedlings showed that an application of 1 g of macerated sugar beet ›Monza‹ prevented sugar beet seeds from germinating and developing. Results could not be obtained from this trial. An application of *H. rhossiliensis* as a fungal suspension significantly reduced root penetration of *H. schachtii* into the roots of sugar beet seedlings. The treatment of root exudates of rape seedlings increased nematode inva-sion into the roots by 30% in comparison to the non-treated control, although the difference was not statistically significant (Figure 39). Nematode invasion into the roots of seedlings treated with Trehalose as a powder was reduced by 32% compared to the control but the difference was not statistically significant. There was also no statictical difference in the number of *H. schachtii* juveniles in the roots of seedlings treated with vanillic acid or the control. Propatox 2%®, on the other hand, signifi-cantly decreased nematode invasion into the roots of treated seedlings by over 70% (Figure 39). In the non-treated linked twin-pot Chamber that served as controls, no significant difference between nematode invasion into the roots of sugar beet seed-lings of either pot was detected.

Figure 39: Effect of Hirsutella rhossiliensis (H.r.), root exudates of rape seedlings (Rape exu-

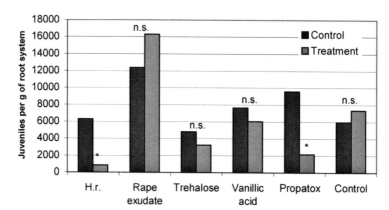

dates), Trehalose as a powder, Vanillic acid as a powder and Propatox 2%® on the number of juveniles of Heterodera schachtii per g of sugar beet root system in second Linked Twin-pot Chamber trials (per g of root system). Means with (*) are significantly different based on T-Test (P ≤ 0.05; n = 5).

8.3.2 Selected additive formulation for trials with Heterodera schachtii and Meloidogyne incognita

Inoculation of pots with an »Empty capsule formulation«, »Baker's yeast + OZ1« or »Capsules with baker's yeast and OZ1« significantly reduced the migration rate of juveniles of *H. schachtii* into treated pots, respectively, in comparison to non-treated pots (Figure 40, Figure 41). The treatments had repellent activity on nematode movement. The amount of nematodes in a pot treated with an »Empty capsule formulation« was 64% less than in a non-treated pot. A treatment of »Baker's yeast with OZ1« led to a reduction of 77% and »Capsules with baker's yeast and OZ1« decreased the amount of nematodes in a treated pot by over 80%.

Figure 40: From left to right: an »Empty capsule formulation«, »Baker's yeast with OZ1« and »Capsules with baker's yeast and OZ1«.

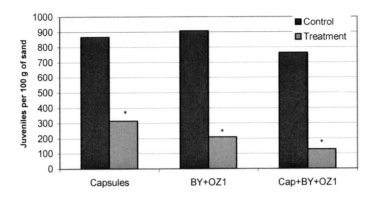

*Figure 41: Effect of an »Empty capsule formulation« (Capsules), »Baker's yeast with OZ1« (BY+OZ1) and »Capsules with baker's yeast and OZ1« (Cap+BY+OZ1) on **the number of Heterodera schachtii juveniles in either treated or non-treated pots** in the Linked Twin-pot Chamber trials (per 100 g of sand). Means with (*) are significantly different based on T-Test (P ≤ 0.05; n = 5).*

The Linked Twin-pot Chamber trial conducted with sugar beet seedlings showed that the additives an »Empty capsule formulation«, »Baker's yeast and OZ1« and »Capsules with baker's yeast and OZ1« did not have attractant activity on juveniles of *Heterodera schachtii* in the additional presence of a seedling. An application of an »Empty capsule formulation« did not have an effect on nematode invasion into the roots of sugar beet seedlings (Figure 42). A treatment of »Baker's yeast and OZ1« reduced the number of nematodes per g of root system by 85% in comparison to the control, although the difference was not statistically significant. An application of »Capsules with baker's yeast and OZ1« significantly decreased nematode invasion by 88% (Figure 42).

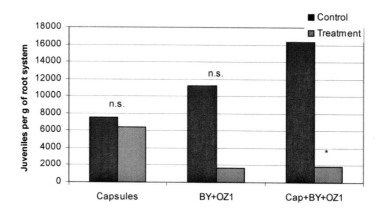

Figure 42: Effect of an »Empty capsule formulation« (Capsules), »Baker's yeast with OZ1« (BY+OZ1) and »Capsules with baker's yeast and OZ1« (Cap+BY+OZ1) on the number of Heterodera schachtii juveniles per gram of sugar beet root system in Linked Twin-pot Chamber trials (per g of root system). Means with (*) are significantly different based on T-Test (P ≤ 0.05; n = 5).

When comparing the Linked Twin-pot Chamber trials conducted with *M. incognita* and *H. schachtii*, respectively, the migration rate of juveniles of *M. incognita* from the tube into either a treated or non-treated pot was 99% less than that of juveniles of *H. schachtii*. Of 5000 inoculated J2 of *M. incognita* per Linked Twin-pot Chamber less than 1% of the nematodes migrated away from the tube and wandered into either the non-treated or treated pot, which makes the analysis of following results questionable. Nonetheless, the treatment »Empty capsules« did not have attractant or repellent activity on J2 of *M. incognita*. An application of »Baker's yeast with OZ1« as a liquid suspension decreased nematode migration into treated pots by over 90%. On the other hand an application of »Capsules containing baker's yeast and OZ1« increased nematode migration by *M. incognita* into treated pots by over 160% (Figure 43). The difference was not statistically significant to the control. When regarding the results, one must take into consideration, that not more than 1% juveniles of *M. incognita* migrated away from the spot of inoculation.

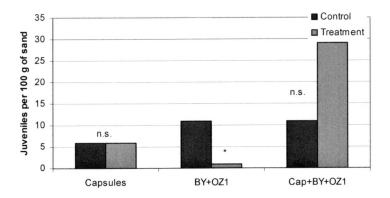

Figure 43: Effect of an »Empty capsule formulation« (Capsules), »Baker's yeast with OZ1« (BY+OZ1) and »Capsules with baker's yeast and OZ1« (Cap+BY+OZ1) on **the number of juveniles of** Meloidogyne incognita **in either treated or non-treated pots** *in Linked Twin-pot Chamber trials (per 100 g of sand). Means with (*) are significantly different based on T-Test (P ≤ 0.05; n = 5).*

Attractant activity was also not detected in the Linked Twin-pot Chamber trials conducted with tomato seedlings ›UC82 Davis‹ and selected materials. An application of empty capsules seemed to have a repellent effect on nematode invasion in comparison to the control, yet the difference was not significant. A treatment of baker's yeast reduced root penetration by nematodes by 82%, which was significantly different to the control (Figure 44). No difference was detected between the number of juveniles in a treated and non-treated root system if »Capsules with baker's yeast and OZ1«(Cap+BY+OZ1) were applied. Furthermore, the trial conducted with the additive »Capsules with baker's yeast and OZ1« (Cap+BY+OZ1) showed the lowest rate of migration from the tube into either the treated or non-treated pot in comparison to trials with »Empty capsules« (Capsules) and »Baker's yeast and OZ1« (BY+OZ1).

105

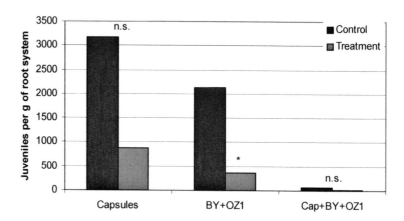

*Figure 44: Effect of an »Empty capsule formulation« (Capsules), »Baker's yeast with OZ1« (BY+OZ1) and »Capsules with baker's yeast and OZ1« (Cap+BY+OZ1) on **the number of juveniles of** Meloidogyne **incognita** per g of tomato root system in Linked Twin-pot Chamber trials (per g of root system). Means with (*) are significantly different based on T-Test (P ≤ 0.05; n = 5).*

8.4 Discussion

Results of the Linked Twin-pot Chamber Trials conducted with juveniles of *H. schachtii* or *M. incognita*, respectively, showed that a high percentage of inoculated nematodes of either genera remained in the tube throughout the duration of the trial. This suggests that either the substrate sand may reduce the efficiency of diffusion of the additive and the capacity of the nematodes to migrate effectively or the selected additives did not act as strong attractants or repellents of nematodes and therefore the effect on nematode migration was only subliminal. In trials conducted without seedlings, the rate of migration of juveniles of *M. incognita* was 99% less than that of *H. schachtii*. The Linked Twin-pot Chamber trial did not prove to be useful for the study of *M. incognita* attraction. Prot and Van Gundy (1981) reported that soil texture appeared to be important for vertical migration of *M. incognita*. They described that the addition of 5% clay to silica sand increased migration markedly over that in pure silicate sand. They assumed that the clay particles added to the silica sand caused migration of root-knot juveniles over large distances to plant roots by holding some root exudates either on clay or in a water film and establishing a gradient which helped nematodes to locate roots. Perhaps the addition of 5% clay to the silica sand could have improved the rate of migration in trials conducted with *M. incognita* and increased the efficiency of diffusion of the added stimuli. Another alternative for improvement of the migration rate of both genera and the efficiency of diffusion of the added stimuli in trials could be the use of a slurry of Sephadex beads as a medium through which nematodes can move easily in an in-vitro assay instead of an in-situ assay conducted in sand. Ward (1973) found that the use of Sephadex led to a very effective assay for attraction to chemical stimuli as it has the advantages that it is chemically more inert, for it contains no charged groups and nematodes can be easily added and removed. Perhaps a nematode attraction assay would have been a good pre-trial for chemo attraction in a solid matrix like sand.

Nonetheless, in the Linked Twin-pot Chamber trials conducted with *H. schachtii* without the presence of plants, a migration rate of over 37% (Trehalose) and up to 85% (Propatox 2%[®]) was detected. None of the additives exhibited any degree of attraction toward juveniles of *H. schachtii*. An application of rapeseed root exudate which is known to induce hatching of eggs within cysts of *H. schachtii* did not have attractant activity on J2 of *Heterodera schachtii* inoculated juveniles. There was no significant affect on nematode migration in comparison to the control. Although Wieser (1956) suggested many years ago, that host roots have repellents as well as attractants, Diez and Dusenbery (1989) found in their studies that root exudates appeared to contain only repellent activity. Extensive efforts to uncover attractant activity by fractionation and other manipulants of the exudates were not successful in revealing attractant activity that was strong or reproducible. An application of 1 g of macerated sugar beet root reduced nematode migration into treated pots by 78% in the first Linked Twin-pot Chamber trial. In the second trial conducted with seedlings of sugar beet ›Dorena‹ the treatment showed phytotoxic effects by inhibiting plant growth of seedlings. It is assumed that the rate chosen for application was too high or that diffusates or metabolites produced during the process of decomposition of macerated sugar beet had phytotoxic effects on seedlings and repulsion towards nematodes. Perhaps different results would have been obtained at lower concentrations of macerated sugar beet root. The application of the alpha linked sugar Trehalose, which acts as a good nutrient for sporulation of *H. rhossiliensis* had no significant effect on juveniles of *H. schachtii* whether sugar beet seedlings were present or not. These results are not unusual because, in 1959, Bird revealed that glucose, fructose, and other sugars which are also root excretions, had no attractive action on two root-knot nematode species. The additive vanillic acid, which was the first structure of a compound with sex pheromone activity in nematodes found by Jaffe et al. (1989), did not attract second-stage juveniles of *H. schachtii*. Most likely this compound only has an attractant effect on the adult male sugar beet cyst nematode such as another sex pheromone of female *Heterodera schachtii* isolated by Jonz et al. 2004 using high performance liquid chromotography. In fact, without the presence of a sugar beet seedling, vanillic acid had a repellent effect on nematode migration. Nematode migration into treated pots was significantly reduced by 44% in comparison to the non-treated pot. Attractant activity and coiling behaviour might be aroused in the male soybean cyst nematode but the effect on second-stage juveniles is repellent without the presence of a seedling. In the presence of a seedling, vanillic acid did not seem to have an effect on nematode orientation and migration to the roots, showing that vanillic acid did not mask host exudates. The migration of nematodes was repelled by an application of Propatox 2%[®] in the Linked Twin-pot Chamber trial conducted without a plant. Without the presence of a plant, an application of Propatox 2%[®] repelled nematode migration by 40%, although the difference was not statistically significant to the control. In the presence of a plant, Propatox 2%[®] reduced nematode penetration into roots of sugar beet seedlings significantly by 88%. Perhaps Propatox 2%[®] masked or neutralized host diffusates, making orientation to roots difficult. The capsule component and endoparasitic fungus *H. rhossiliensis* did not influence nematode migration in the trial conducted without a seedling but in the trial conducted with a seedling, nematode penetration was significantly reduced by an application with *H. rhossiliensis*. As this trial was harvested four days later than the trial without a plant, there is a great possibility that *H. rhossiliensis* was able to sporulate, infect and parasitize juveniles of *H. schachtii*, thus reducing nematode invasion of roots significantly in comparison to the control.

The initial idea to add an additive to the capsule formulation of *H. rhossiliensis* which might act as an attractant of nematodes and thus improve infection of nematodes with spores of *H. rhossiliensis* did not succeed. The six additives selected did not possess attractant activity, macerated sugar beet and Propatox 2%[®] even had a repulsive effect on nematode behaviour. Such additives would reduce the efficacy of the capsule. The addition of Trehalose as a nutrient for the sporulation of *H. rhossiliensis* should not be a cause for concern as it has neither attractant nor repellent activity on juveniles of *H. schachtii*.

Linked Twin-pot Chamber trials conducted with the additive »Baker's yeast and OZ1«, »Capsules with baker's yeast and OZ1« and »Empty capsules« exhibited interesting results on migration and root penetration of juveniles of *H. schachtii*. Trials conducted with *M. incognita* with or without tomato seedlings showed that less than 1% of the nematodes extracted from the trial migrated away from the spot of inoculation. Although the additive »Baker's yeast and OZ1« had a significant repellent effect on less than 1% of juveniles that left the spot of inoculation, whether the trial was conducted with a tomato seedling or not, one cannot speak of nematode repulsion at such a low rate of nematode migration. Nematode migration in first Linked Twin-pot Chamber trials conducted with *H. schachtii* without the presence of plants was much higher. The migration rate was over 66% in the trial conducted with empty capsules. It was 80% if »Baker's yeast and OZ1« was applied and 52% after an application of capsules with »Baker's yeast and OZ1«. Each application led to a significant repellence of juveniles of *H. schachtii*. In second Linked Twin-pot Chamber trials conducted with *H. schachtii* only the treatment »Capsules with baker's yeast and OZ1« repelled migration into treated pots significantly in comparison to the control. Perhaps the repellence of an »Empty capsule formulation« and »Baker's yeast and OZ1« was too weak to be effective if another stimulus, such as root exudates of *H. schachtii* seedlings, with a stronger attraction was present. Such a stimulus could be carbon dioxide, which is quantitively the most important of root excretions (Klinger, J. 1965). It is produced during root respiration. Observations made by other researchers have suggested that plant-parasitic nematodes are attracted to carbon dioxide and that carbon dioxide is the principle means by which nematodes locate hosts (Klingler, 1963; Pline and Dusenbery, 1987; Prot, 1980). »Capsules with baker's yeast and OZ1«, on the other hand, did significantly reduce root penetration of *H. schachtii* per g of root system in comparison to the non-treated control. The individual components of the capsule did not repell nematodes significantly but in combination an effect was detected. The combination of both seems to have a synergistic effect on repellent activity. The trial has to be repeated to verify such findings. If these results can be confirmed, then such a capsule type without the addition of the endoparasitic fungus *H. rhossiliensis* could be incorporated into the soil in the area of the growing plant in order to repel nematodes and protect plant roots from nematode invasion. Nonetheless, such a capsule type would not be effective in reducing the nematode population density.

The trials conducted have shown that none of the additives tested had attractant activity on nematodes. The inclusion of any of the selected additives into the capsule formulation would not be a benefit to the formulation in terms of attracting nematodes to the capsule for a better infestation of juveniles with spores of *H. rhossiliensis*.

8.5 Summary

Various substances were tested in Linked Twin-pot Chamber trials, conducted with and without seedlings for their attractant activity towards infective juveniles of *Heterodera schachtii* and *Meloidogyne incognita*. Our goal was to discover a potential attractant that could be applied as an additive to the capsule formulation of the endoparasitic fungus *H. rhossiliensis* in order to improve the infestation rate of infective juveniles. None of the substances tested resulted in attractant activity. Among the substances tested, the substances macerated sugar beet, vanillic acid and Propatox 2%® and Capsules containing »OZ1 and Baker's yeast« tested on *H. schachtii* even revealed repellent activity.

9 Liquid and solid formulations as alternatives to encapsulation of *Hirsutella rhossiliensis*

9.1 Introduction

So far, our research was devoted to the biological control effectiveness of encapsulation of the endoparasitic fungus *H. rhossiliensis* using derivates of guargum, pectine and alginate against the plant-parasitic nematode species *Heterodera schachtii* and *Meloidogyne incognita*. The effectiveness of the different types of encapsulation containing vegetative hyphae of *H. rhossiliensis* was insufficient if capsules were applied to non-treated soil. The fungus within the capsule was exposed to biotic inhibition and failed to grow. Jaffee (2000) also found that the fungus was more sensitive to biotic inhibition when formulated as encapsulated vegetative hyphae but that it was insensitive when formulated as parasitized nematodes. Given the importance of assimilative hyphae in sporulation of *H. rhossiliensis* and the success obtained with colonized nematodes as inoculum, Lackey et al. (1992) hypothesized that nematodes could also be parasitized by artificially infesting soil with fungal hyphae. He demonstrated that hyphae similar to those in the nematode could be grown in a nutrient solution and sporulate when removed from the solution and added to soil. Based on his results, the biological control potential of the fungus as vegetative hyphae, grown in liquid shake culture and suspended in a liquid formulation or cultured on solid media was evaluated in the present studies. Selected liquid formulations contained organic substrates for the fungus to utilize in order to improve its effectiveness against the sugar beet cyst nematode. Another effort to improve the efficacy of a liquid formulation was made by adding juveniles of *H. schachtii* to liquid shake culture of *H. rhossiliensis* in the attempt to stimulate the fungus and perhaps increase its virulence through the presence of its host during liquid shake culture. Besides liquid formulations, also solid formulations as alternatives to an encapsulation of *H. rhossiliensis* were chosen. Nonetheless, like an encapsulation, other substrates such as wheat seeds and corn grits do not protect the germ tube of the fungus from antagonism like a host nematode does (Jaffee and Zehr, 1985). An effort to overcome biotic inhibition of the fungus was to let the fungus grow from these substrates, produce aerial mycelia and sporulate prior to application in soil. Solid formulations were a wheat seed culture, a corn grit culture of *H. rhossiliensis* and *H. rhossiliensis* coated on crystal quartz gravel. The efficacy of the liquid and solid formulations was tested in a series of greenhouse trials against *H. schachtii* on sugar beet.

9.2 Material and Methods

9.2.1 Production of liquid formulations

The effectiveness of liquid formulations of *H. rhossiliensis* was tested against *H. schachtii* on sugar beet. For the following studies four different liquid formulations were chosen. The differentiation of the first three formulations was in the nutrient composition, in which vegetative hyphae of *H. rhossiliensis* was suspended. The basic preparation for all four liquid formulations was identical. For this, vegetative hyphae were cultured in a nutrient solution as described in chapter 2.3.5 »Fermentation and encapsulation of *Hirsutella rhossiliensis*«. Hereafter, the fungal pellets, consisting of vegetative hyphae, were collected on a 500 µm aperture sieve (INOX 18/10; Fackelmann GmbH&Co. KG; Hersbrück, Germany) and rinsed three times with sterile distilled water. Based upon techniques described by Liu and Chen (2005), 20 g of collected, moist, vegetative hyphae of *H. rhossiliensis* were added to a glass beaker filled with 250 ml of a sterile liquid solution for each liquid formulation, which had been autoclaved for 20 minutes at 121°C. The mixture was blended with a kitchen hand mixer (Alaska®, SIM 120, Germany) for approximately 2 minutes with intervals until a homogenous suspension was obtained. For the first liquid formulation (**H.r. + Water**), 20 g of vegetative hyphae were suspended into 250 ml of sterile distilled water. For the second liquid formulation (**H.r. + Nutrients A**), 20 g of vegetative hyphae were suspended in 250 ml of a PDA solution diluted to 20% to prevent the solution from coagulating at room temperature. The third liquid formulation (**H.r. + Nutrients B**) consisted of 20 g of vegetative hyphae which had been suspended in 250 ml of a nutrient solution consisting of Glucose (5 g), Yeast extract (1.25 g), MES-NaOH buffer (1 g) and K_2HPO_4 (0.25 g).

Besides adding nutrients to the fungal suspension, another effort to improve the efficacy of liquid formulations was made by adding juveniles of H. schachtii to liquid shake culture of H. rhossiliensis in the attempt to stimulate the fungus and perhaps increase its virulence through the presence of its host during liquid shake culture. The preparation of the liquid culture of H. rhossiliensis was as described in chapter 2.3.5 »Fermentation and encapsulation of H. rhossiliensis«. A variation was made by adding 500,000 surface sterilized juveniles of H. schachtii in 1 ml of sterile distilled water to a flask filled with 500 ml of liquid culture of H. rhossiliensis. For sterilization, second-stage juveniles of H. schachtii were placed in a bath of 0.02% mercury chloride solution, followed by three successive one minute baths of sterile distilled water. After twelve days, vegetative hyphae were collected and rinsed. For the fourth liquid formulation (H. r. (H. s.) + Water), 20 g were suspended into 250 ml of sterile distilled water as described above. Each liquid formulation was tested for contamination by plating 100 µl of the liquid formulation on PDA plates. Furthermore, the number of colony forming units per ml of fungal suspension was assessed.

9.2.2 Production of solid formulations

Solid culture on wheat seeds

Wheat seeds were coated with vegetative hyphae, which sporulated under sterile conditions prior to application in the soil. The preparation of the solid wheat culture was as following: Wheat seeds with a given seed moisture of 12% were placed into a Bosch[®] kitchen mixer (Robert Bosch AG, Stuttgart, Germany). To obtain a seed moisture of 30%, the predetermined amount of water was added to the seeds in the blender. The wheat seeds and water were mixed for about two minutes with intervals until all the wheat seeds had been chopped thoroughly. One-hundred grams of moist, chopped wheat seeds were filled into each of three 250 ml glass Schott[®] bottles (Schott AG, Mainz, Germany) and autoclaved at 120°C for 90 min. Hereafter, three fungal suspensions, consisting of 10 g of shredded vegetative hyphae of *H. rhossiliensis*, suspended in an autoclaved solution of 40 ml of water and 0.4 g of PDA powder, were filled into the three bottles each containing 100 g (=200 ml) of autoclaved chopped wheat seeds. The bottles were placed on a roller (RM5, Witeg Electrics Labortechnik GmbH, Wertheim, Germany) for an hour. Then the content of each bottle was poured into one of three glass 1l Erlenmeyer flasks (Fisher Scientific GmbH, Schwerte, Germany). The flasks were placed in an incubator (Sanyo Fisher Sales GmbH, München, Germany) at 23°C for six weeks to allow optimal growth of aerial mycelia and sporulation of *H. rhossiliensis*. Contamination was checked throughout incubation. First growth of aerial mycelia was visible after four days.

Solid culture on corn grits

Two solid cultures of corn grits were prepared according to Liu and Chen (2001). For this purpose corn grits, silica sand and water were mixed by weight at a ratio of 2:2:1 and autoclaved two times at 120°C for 20 min. Then, 250 g (= 200 ml) of corn grits were filled into each of 250 ml sterile glass Schott[®] bottles. Hereafter, the corn grits were treated with 20 ml of fungal suspension, containing 1.6 g of fresh vegetative hyphae of *H. rhossiliensis*. The content of the bottle was shaken to allow for a homogenous mixture of the components. The inoculated corn grits of each Schott® bottle were then filled into sterile glass dishes (ø = 20 cm; height = 2.5 cm). The dishes were incubated for six weeks at 23°C. Once every two weeks a sample was collected and examined for contamination.

Coating of crystal quartz gravel[®]

Among formulations chosen for *H. rhossiliensis*, a novel type of formulation was introduced in this study. As organic substrates were attractive for antagonism, a solid carrier was chosen for the fungus that did not act as a nutrient source for other microorganisms, which was crystal quartz gravel[®] (Dennerle Company, Vinningen, Germany). Crystal quartz gravel[®] is free of hardness builders, neutral in water, CO_2-resistant, with rounded off edges and a grain size of 1 – 2 mm. The gravel was rinsed with warm water three times before use in order to remove debris and subsequently dried on a kitchen towel. Hereafter, two 250 ml glass Schott® bottles were each filled with 300 g (=200 ml) of rinsed crystal quartz gravel® and autoclaved at 121°C for 40 minutes for sterilization. A PDA solution was prepared by dissolving 15 g of PDA powder into 250 ml of sterile distilled water and autoclaved at 121°C for 40 minutes. Under sterile conditions, the hot PDA solution was poured into the glass Schott® bottles with crystal quartz gravel® until the gravel was covered with PDA solution, then a lid was held at the opening to prevent gravel from falling out of the bottle while excess PDA solution was poured out of the bottle. The purpose was to coat the crystal quartz gravel® with PDA, not embed it in it. The crystal quartz gravel® coated with PDA was poured into sterile glass dishes with a diameter of 20 cm and 2.5 cm in height and left on a clean bench (Biowizard Kojair® KR-130 BW, Kojair Tech Oy, Vilppula, Finnland) to dry. After the PDA hardened, the gravel was inoculated with spores of *H. rhossiliensis*. For this purpose, two PDA plates with 14 day-old cultures of sporulating *H. rhossiliensis* were needed for the inoculation of coated gravel per glass dish. Square pieces of PDA, approximately 1.5 cm^2 in size, containing sporulating aerial mycelia of *H. rhossiliensis* were rubbed against the coated gravel of each glass dish. A glass lid was placed on the dish to prevent contamination and the dishes were placed in a Sanyo® incubator at 23°C for six weeks. During this period, samples were taken to examine for possible contaminations.

9.2.3 Comparison of liquid formulations of Hirsutella rhossiliensis for control of Heterodera schachtii on sugar beet

A 100 ml biotest was conducted to determine the efficacy of the liquid formulations 1-3 (H.r. + Water, H.r. + Nutrients A, H.r. + Nutrients B) described in chapter 9.2.1 »Production of liquid formulations«. The effectiveness of these three liquid formulations was investigated against the sugar beet cyst nematode *Heterodera schachtii* on sugar beet ›Dorena‹.

Treatments per 100 ml of soil:
(1) Non-treated control (3.3 ml of sterile water)
(2) Nutrients A (3.3 ml of a sterile 20% PDA solution)
(3) Nutrients B (3.3 ml of a sterile nutrient solution, see chapter 9.2.1 »Production of liquid formulations« for composition)
(4) H.r. + Water (0.26 g of *H. rhossiliensis* suspended in 3.3 ml of sterile water)
(5) H.r. + Nutrients A (0.26 g of *H. rhossiliensis* suspended in 3.3 ml of a sterile 20% PDA solution)
(6) H.r. + Nutrients B (0.26 g of *H. rhossiliensis* suspended in 3.3 ml of a sterile nutrient solution, see chapter 9.2.1 »Production of liquid formulations« for composition)

The trial was conducted in a field soil/sand mixture (2:1; w/w) with a pH of 6.2. The substrate was heat-treated at 180°C for 3 hours prior to trial set-up. To obtain a population density of 1500 eggs and juveniles of *H. schachtii* per 100 ml of soil, 10 l of soil were mixed thoroughly with 80 g of cyst inoculum (see chapter 2.? for details). Soil in 100 ml lots, were placed in plastic bags. Liquid formulations with and without *H. rhossiliensis* were added, then the contents of the bag were mixed thoroughly for a homogenous distribution. The treated lots were filled into 100 ml plastic containers (40 x 20 x 120 mm, Kelder Plastibox b. v., s'-Heerenberg, The Netherlands). Twelve replicates were used. The plastic containers were placed in the greenhouse at 20°C ± 3°C and were watered as needed. Two weeks after trial set-up, sugar beet seeds ›Dorena‹ were sown into the soil of each container. Seven days after germination sugar beet seedlings were removed and the number of juveniles of *H. schachtii* that penetrated the root system was counted by staining the roots with a 0.02% acid fuchsin solution according to Byrd et al. (1983).

9.2.4 Efficacy of Hirsutella rhossiliensis *as a liquid suspension or as a solid formulation*

The efficacy of liquid formulations »H.r. + water« and »H.r. (H.s.) + water« described in chapter 9.2.1 »Production of liquid formulations« was tested in two 100 ml biotests against *H. schachtii* on sugar beet ›Dorena‹. The two formulations differed in the liquid culture of the fungus »*H. rhossiliensis*« to which 500,000 juveniles of *H. schachtii* were added to 500 ml of liquid culture of *H. rhossiliensis* for stimulation of the fungus during fermentation. Furthermore, the effectiveness of a solid culture of *H. rhossiliensis* on wheat seeds was investigated (see chapter 9.2.2 »Production of solid formulations«) in comparison to liquid formulations. Also, the influence of application time on the efficacy of these formulations was examined in two biotests.

Treatments of biotest I and biotest II per 100 ml of soil:
  (1) Non-treated control (CON) (3.3 ml of sterile water)
  (2) H.r. + Water (3.3 ml)
  (3) H.r. (H.s.) + Water (3.3 ml)
  (4) Wheat seeds (1 g)
  (5) Wheat seeds + H.r. (1 g)

Both 100 ml biotests were conducted in a field soil and sand mixture (2:1; w/w) with a pH of 6.2. The substrate was heat-treated at 180°C for 3 hours prior to trial set-up. To obtain a population density of 1500 eggs and juveniles of *H. schachtii* per 100 ml of soil, 15 l of soil were mixed thoroughly with 120 g of cyst inoculum soil (see chapter 2.2.1 for details). One-hundred ml lots of soil containing cysts were placed into plastic bags. Treatments of the liquid formulations »H.r. + Water« and »H.r. (H.s.) + Water« at 0.26 g of mycelium suspended in 3.3 ml of water or 1 g of a solid culture of *H. rhossiliensis* on wheat seeds were added and mixed thoroughly. Controls consisted of soil amended with 3.3 ml of water or 1 g of non-treated wheat seeds. The lots of treated soil were filled into 100 ml plastic containers. Each treatment had twelve replicates. Trial set-up for both biotests was identical. The biotests were set in the greenhouse at 20°C ± 3°C. Irrigation was added as needed. Two sugar beet seeds ›Dorena‹ were sown in each container of biotest I ten days after trials set-up and in biotest II seventeen days after trial set-up. Seven days after germination, the number of juveniles of *H. schachtii* within a root system was determined after staining the juveniles with an acid fuchsin solution (Byrd et al. 1983).

9.2.5 Efficacy of liquid and solid formulations

The efficacy of the liquid and solid formulations of *H. rhossiliensis* produced and introduced in this chapter was tested against *H. schachtii* on sugar beet in a 300 ml pot trial. The treatments were as follows:

Treatments of the pot trial per 300 ml of soil:
- (1) Non-treated control 1 (10 ml of water)
- (2) Non-treated control 2 (10 ml of water)
- (3) H.r. + Water (10 ml)
- (4) H.r. (H. s.) + Water (10 ml)
- (5) Nutrients A (10 ml)
- (6) H.r. + Nutrients A (10 ml)
- (7) Nutrients B (10 ml)
- (8) H.r. + Nutrients B (10 ml)
- (9) Crystal quartz gravel (9 g = 6 ml)
- (10) Crystal quartz gravel + H.r. (9 g = 6 ml)
- (11) Wheat seeds (3 g = 6 ml)
- (12) Wheat seeds + H.r. (3 g = 6 ml)
- (13) Corn grits (4.5 g = 6 ml)
- (14) Corn grits + H.r. (4.5 g = 6 ml)

The 300 ml pot trial was conducted in heat-treated field soil-silicate sand mixture (2:1, w/w). The soil substrate had pH 6.2. At day 0, 300 ml of soil were thoroughly mixed with the treatment and 2.4 g of cyst inoculum containing 4,500 eggs and juveniles of *H. schachtii* (see chapter 2.2.1 Preparation of cyst inoculum of *Heterodera schachtii*) and placed back into a 300 ml pot. Each treatment had 12 replicates. The pots were set in the greenhouse at an average temperature of 20°C +/- 3°C. Irrigation was applied as needed. On day 12, two sugar beet seeds ›Dorena‹ were sown into each pot. After germination seedlings were thinned to one per pot. Soil temperature was measured to determine the completion of a life-cycle of *H. schachtii* by calculation of temperature day degrees (Greco et al., 1982). Three-hundred day degrees above a base of 10°C are required for completion of a generation, which was achieved after seven and a half weeks when the experiment was terminated. Plant shoot fresh weight was recorded and cysts of *H. schachtii* were extracted from a soil subsample of 200 g according to Jenkins (1964), counted and the number of eggs and juveniles within the cysts were recorded.

9.3 Results

9.3.1 Comparison of liquid formulations of Hirsutella rhossiliensis for control of Heterodera schachtii on sugar beet

The number of colony forming units (cfu) per ml of liquid formulation was similar for each of the three formulations. The numbers ranged from $1.1*10^6$ (H.r. + Water) to $1.5*10^6$ cfu/ml (H.r. + Nutrients B). Furthermore, vitality assays proved that the tested formulations were not contaminated by other microorganisms.

Results of the 100 ml biotest conducted with different liquid formulations showed that liquid formulations without *H. rhossiliensis* did not significantly reduce the number of juveniles per g of root system in comparison to the non-treated control (Water (-)) (Figure 45). All three liquid formulations containing *H. rhossiliensis* reduced nematode invasion by over 53%, in comparison to the non-treated control (Water (-)). The differences were statistically significant On the other hand, additional nutrients added to the liquid formulations did not increase the efficacy of *H. rhossiliensis*. The liquid formulations with *H. rhossiliensis* containing nutrients (H.r.+Nutrients A, H.r.+Nutrients B) did not improve the efficacy of *H. rhossiliensis* when compared to liquid formulation »H.r. + water«, which consisted of only water and *H. rhossiliensis*.

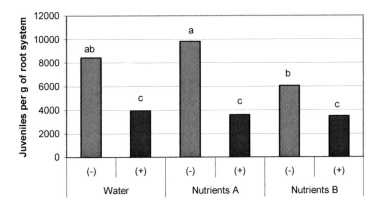

Figure 45: Effect of different liquid formulations containing Hirsutella rhossiliensis *applied to heat-treated soil on nematode invasion of* Heterodera schachtii *in the roots of sugar beet seedlings. Water = liquid formulation consisting of water, Nutrients A = liquid formulation consisting of water and PDA powder, Nutrients B = liquid formulation consisting of water, yeast, glucose, MES-NaOH buffer, and K_2HPO_4, (-) = Liquid formulation without shredded, vegetative hyphae of H. rhossiliensis, (+) = Liquid formulation with shredded, vegetative hyphae of H. rhossiliensis. Values are means of 12 replicate plastic containers; means followed by the same letter are not significantly different according to Tukey HSD with $p \leq 0.05$.*

9.3.2 Efficacy of Hirsutella rhossiliensis *as a liquid suspension or as a solid formulation*

The number of colony forming units for the liquid formulation »H.r. + Water«, to which nematodes were not added during liquid culture, was $5.5 * 10^5$ cfu/ml, whereas the liquid formulation »H.r. (H.s.) + Water«, to which 500,000 J2 of *H. schachtii* were added during liquid culture, had a cfu/ml of $4.2 * 10^5$ (see chapter 9.2.1 »Production of liquid formulations« for details). The liquid formulations were not contaminated by other microorganisms. Results of vitality assays conducted with a solid culture of *H. rhossiliensis* on wheat seeds were also free of contamination. Aerial mycelia of *H. rhossiliensis* successfully grew from the wheat seeds.

The application time played a significant role on the efficacy of *H. rhossiliensis* against *H. schachtii*. After ten days of exposure to treatments »H.r. (H.s.) + Water« and wheat seeds coated with aerial mycelia of *H. rhossiliensis* (Wheat seeds + H. r.) nematode invasion of *H. schachtii* juveniles into the roots of sugar beet seedlings was reduced by 30% and 40%, respectively. The differences, although, were not statistically significant in comparison to the non-treated control (Figure 46). Wheat seeds without *H. rhossiliensis* did not reduce nematode invasion. Extending the exposure time to 17 days increased the efficacy of *H. rhossiliensis* as a liquid and solid formulation. Both liquid formulations of *H. rhossiliensis* (H.r. + Water and H.r. (H.s.) + Water) led to a better reduction of root penetration of 43% and 62%, respectively, whereas the difference caused by »H.r. (H.s.) + Water« was statistically significant to the control (Figure 47). An exposure time of 17 days to a treatment of a solid culture of *H. rhossiliensis* on wheat seeds led to a higher decrease of nematode invasion into the roots of sugar beet seedlings. Nematode invasion was significantly reduced by 84% in comparison to the control (Figure 47).

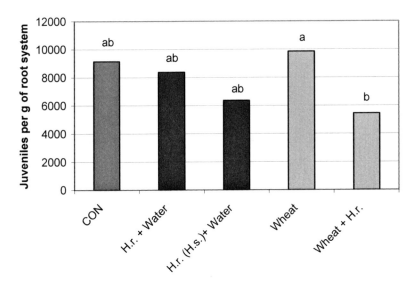

Figure 46: Effect of a fungal suspension and a solid formulation of Hirsutella rhossiliensis *applied to heat-treated soil on nematode invasion of* Heterodera schachtii *in the roots of sugar beet seedlings after* **10 days** *of exposure time. CON = Non-treated control, H. r. + water = Liquid formulation consisting of water and shredded vegetative hyphae of* H. rhossiliensis, *H.r. (H.s.) + water = Liquid formulation consisting of water and shredded vegetative hyphae of* H. rhossiliensis *(to which 500,000 second-stage juveniles of* H. schachtii *were added during liquid culture of the fungus and removed after collection of vegetative hyphae of* H. rhossiliensis), *Wheat = wheat seeds containing a seed moisture of 30%, Wheat + H.r. = Solid culture of* H. rhossiliensis *on wheat seeds. Values are means of 12 replicate plastic containers; means followed by the same letter are not significantly different according to Tukey HSD with p ≤ 0.05.*

119

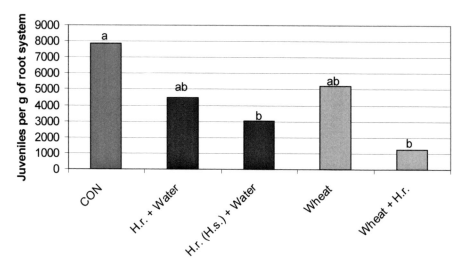

Figure 47: Effect of a fungal suspension and a solid formulation of Hirsutella rhossiliensis *applied to heat-treated soil on nematode invasion of* Heterodera schachtii *in the roots of sugar beet seedlings after **17 days** of exposure time. CON = Non-treated control, H.r. + Water = liquid formulation consisting of water and shredded vegetative hyphae of* H. rhossiliensis, H.r. (H.s.) + Water = Liquid formulation consisting of water and shredded vegetative hyphae of H. rhossiliensis (to which 500,000 second-stage juveniles of H. schachtii were added during liquid culture of the fungus and removed after collection of vegetative hyphae of H. rhossiliensis), Wheat = wheat seeds containing a seed moisture of 30%, Wheat + H.r. = Solid culture of H. rhossiliensis on wheat seeds. Values are means of 12 replicate plastic containers; means followed by the same letter are not significantly different according to Tukey HSD with p ≤ 0.05.*

9.3.3 Efficacy of liquid and solid formulations

In the 300 ml pot trial conducted with various liquid and solid formulations, re-sults showed that none of the treatments applied, significantly reduced the number of eggs and juveniles of H. schachtii per 100 g of soil after harvest in comparison to the non-treated controls. Regardless, whether the formulation contained the fungus H. rhossiliensis or acted as an empty formulation, the final population density was not be decreased significantly. Furthermore, a weak positive correlation between the variables plant shoot fresh weight in g and the number of eggs and juveniles per 100 g of soil after harvest (CSpearman's rho = 0.504; p ≤ 0.001, SPSS 11.0) was de-tected. The graphic below displays the number of eggs and juveniles per 100 g of soil and the plant shoot fresh weight in g (Figure 48). In general, a low plant shoot fresh weight of sugar beet seedlings resulted in a lower number of eggs and juveniles per 100 g of soil with some exceptions. Plant shoot fresh weight of sugar beet seedlings treated with liquid formulations »H.r. + Water« and »H.r. (H.s.) + Water« could be compared to that of the non-treated control (CON) no. 1, whereas an application of the liquid formulation »H.r. + Water« increased nematode reproduction by 45% in comparison to the non-treated control. The application of the liquid formulation »H.r. (H.s.) + water« had no effect on nematode reproduction. Nonetheless, both differences were not significant in comparison to control no. 1. A treatment of the liquid formulation »Nutrients A« without the addition of *H. rhossiliensis* (-) led to a reduction reduction of plant shoot fresh weight of 62% and a reduction of the final population density by 35% in comparison to control no. 1. On the contrary, this formulation with the addition of *H. rhossiliensis* increased plant shoot fresh weight by 29% but also led to an increase of eggs and juveniles per 100 g of soil at harvest. The differences were not statistically significant in comparison to the non-treated control no. 1. Results obtained by an application of »Nutrients B« without the addition of *H. rhossiliensis* did not differ strongly from the control. The addition of *H. rhossiliensis* to this liquid formulation increased plant shoot weight and also the number of eggs and juveniles per 100 g of soil, although the increase was not statistically significant to the control. The solid formulation of Crystal quartz gravel® with and without the addition of *H. rhossiliensis* reduced plant shoot weight by approximately 43% in comparison to the control, whereas the addition of the fungus reduced the amount of eggs and juveniles per 100 g of soil by 28% and the treatment without fungus had no effect on nematode reproduction in comparison to non-treated control (CON) no. 1. An application of wheat seeds with and without fungus had no effect on plant shoot weight but nematode reproduction was reduced by over 30%, whereas the treatment without *H. rhossiliensis* led to a reduction of 50% in comparison to the non-treated control. The differences were not statistically significant to the controls. An application of corn grits with the fungus (Corn grits (+)) did not have an effect on plant shoot weight or nematode reproduction. Yet, without the fungus, a reduction over 75% was obtained for plant shoot weight as well as nematode reproduction (Figure 47).

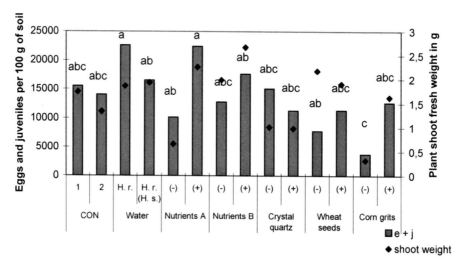

Figure 48: The effect of liquid and solid formulations with and without Hirsutella rhossiliensis *on plant shoot fresh weight of sugar beet seedlings at harvest and the number of eggs and juveniles per 100 g of soil after harvest. CON 1/2 = Non-treated control no. 1 and 2, H.r. + Water = Liquid formulation consisting of water and shredded vegetative hyphae of* H. rhossiliensis, H.r. *(H.s.) + Water = Liquid formulation consisting of water and shredded vegetative hyphae of* H. rhossiliensis *(to which 500,000 second-stage juveniles of* Heterodera schachtii *were added during liquid culture of the fungus and removed after collection of vegetative hyphae of* H. rhossiliensis), (-) = formulation without *H.* rhossiliensis, (+) = formulation with *H.* rhossiliensis, *Nutrients A = liquid formulation consisting of water and potato dextrose agar powder, Nutrients B = liquid formulation consisting of water, yeast, glucose, MES-NaOH buffer, K₂HPO4. Values are means of 12 replicate 300 ml pots; means followed by the same letter are not significantly different according to Tukey HSD with p ≤ 0.05.*

9.4 Discussion

The 100 ml biotest conducted in heat-treated soil to compare the efficacy of three different liquid formulations (H.r. + Water, H.r. + Nutrients A, H.r. + Nutrients B) demonstrated that each formulation was similarly effective in suppressing penetration of sugar beet roots by *H. schachtii*. The addition of nutrients did not significantly increase the efficacy of a liquid formulation containing *H. rhossiliensis*. Nematode invasion was reduced by approximately 50% by each fungal suspension of *H. rhossiliensis*. The idea that supplemental nutrients added to liquid formulations of *H. rhossiliensis* might increase sporulation of the fungus and improve the efficacy of the fungus in suppressing root penetration by *H. schachtii* could not be verified in this trial. Jaffee (2000) described that sporulation involved converting previously assimilated food (in liquid shake culture) into new hyphae, phialides, and conidia and that added nutrients were not required. Furthermore, if applied in non-treated soil, more aggressive saprophytes might make use of the added nutrients and perhaps attack or otherwise inhibit *H. rhossiliensis*. Our attempt to improve the formulation did not succeed, most probably due to the fact that supplemental nutrients were not required. Many attempts have been made to improve the efficacy of *H. rhossiliensis* by developing novel formulations but it has been difficult to achieve until now. Lackey et al. 1993 reported a reduction of nematode invasion in roots of tomato seedlings by *M. javanica* by 50% when 50 alginate pellets of *H. rhossiliensis* were added to 100 cm^3 of loessy sand infested with egg masses of the nematode. Tedford et al. 1995 found that 50% suppression of root penetration required 0.9 alginate pellets with fungus per cm^3 of soil if *H. schachtii* were placed 2 cm from the roots and 0.3 pellets per cm^3 of soil if nematodes were placed 4 cm from the roots. The fungal suspensions applied in the 100 ml biotest are mixed thoroughly throughout the soil and had the same efficacy as alginate capsules applied by Lackey et al. (1993) and Tedford et al. (1995), which are not as finely distributed as a fungal suspension.

The following 100 ml biotest conducted in heat-treated soil with two different fungal suspensions of H. rhossiliensis and coated wheat seeds of H. rhossiliensis demonstrated that the period of treatment of the fungus played a role in the efficacy of the formulation containing the fungus. An application of formulations containing H. rhossiliensis ten days before sowing sugar beet seeds were much less effective than applications applied 17 days before sowing. A low efficacy of the liquid formulations containing H. rhossiliensis was expected after merely 10 days for it is known that sporulation from parasitized H. schachtii requires 14 days at 20°C, after 14 days the nematode substrate has been depleted and no new spores are produced (Jaffee et al., 1990). As vegetative hyphae are comparable to parasitized H. schachtii, the same amount of time is expected for vegetative hyphae of H. rhossiliensis to sporulate. Thus exceptional results were obtained by a treatment of H.r. (H.s.) + water containing H. rhossiliensis, to which 500,000 J2 of H. schachtii were added during liquid culture. A reduction of root penetration of H. schachtii in roots of sugar beet seedlings by 30% was achieved in comparison to the non-treated control, whereas »H.r. + Water« reduced nematode invasion by merely 8% in comparison to the non-treated control. Both results did not differ significantly from the control. A treatment of »H.r (H.s.) + water« 17 days prior to seeding led to a significant suppression of root penetration by 61.5%, in comparison to 43% achieved by the liquid formulation »H. r. + Water«. Nutrients added to the liquid formulation had no influence on the efficacy of the fungal formulation as demonstrated in the previous trial, whereas the addition of J2 of H. schachtii to liquid culture shows greater potential. Our hypothesis of adding juveniles of H. schachtii to the liquid culture of H. rhossiliensis as a potential stimulant had an effect on the efficacy of the formulation. To which extent it has an effect needs to be studied in further in-vitro experiments to determine whether sporulation is accelerated or whether the virulence of the strain is improved by the addition of H. schachtii juveniles during liquid culture of the fungus. A treatment of wheat seeds coated with aerial mycelia of H. rhossiliensis ten days prior to sowing of seeds achieved a reduction of root penetration by 40%, whereas uncoated seeds had no effect in comparison to the non-treated control. As H. rhossiliensis was sporulating when it was applied to heat-treated soil, a higher efficacy of the fungus was expected. Better results were obtained if applied 17 days prior to sowing, then the suppression of root invasion was 84% in comparison to the non-treated control, although uncoated wheat seeds also suppressed nematode invasion by 34%. Perhaps the application time plays an important role, because spores of H. rhossiliensis stick to the soil after being thoroughly mixed into the soil and the fungus needs time to regenerate. The suppression of nematode invasion by 84% is far better than the results obtained by liquid formulations. Though, it was noticeable that coated wheat seeds lying on the surface of the soil, which were exposed directly to air and water, were infested with mold. As soon as the wheat seeds are exposed to non-sterile surroundings they attract secondary contaminants. Jaffee and Zehr (1985) also stated the good germination of H. rhossiliensis on wheat seeds under sterile conditions. Unfortunately, in non-treated soil, growth of H. rhossiliensis was greatly suppressed. Initially, our intension was to have aerial mycelia successfully cover the entire seed, thus giving H. rhossiliensis an advantage in growth towards other competitive microorganisms but the fungus seems to be such a weak saprophyte that even if aerial mycelia of H. rhossiliensis fully coats the entire seed, it still lacks competitiveness.

Liquid and solid formulations of H. rhossiliensis applied in the 300 ml pot trial conducted in heat-treated soil on sugar beet ›Dorena‹ against H. schachtii were not effective in reducing the final the nematode population density after one life-cycle of H. schachtii. Although the liquid formulations were all previously tested in 100 ml biotests and proved good results in suppressing nematode invasion of H. schachtii into the roots of sugar beet seedlings, the same formulations did not reduce nematode numbers in the 300 ml pot trial. The fungal liquid formulations at an application rate of 0.8 g mycelium/per pot did not control nematodes. Liu and Chen (2005) obtained different results in their 300 ml pot trial conducted in heat-treated soil. Their fungal suspension, also applied at a rate of 0.8 g mycelium/per pot achieved a reduction of 80% in the number of eggs/ml of Heterodera glycines in comparison to the non-treated control. Initially, for our 300 ml pot trial, the application rate of 0.8 g/per pot of mycelia of H. rhossiliensis was chosen based on the results of Liu and Chen (2005). Additionally, by adding nutrient components to the formulations or J2 of H. schachtii to the liquid culture of H. rhossiliensis, we had hoped to achieve a more effective control than Liu and Chen (2005). Since applications of fungal liquid suspensions containing Nutrients A or B managed to, at least, increase plant shoot fresh weight in comparison to the control, it was assumed that plants were perhaps protected to a certain extent within the first few weeks of vegetation. However, the rich supply of nutrients allows females to mature into cysts containing a great number of eggs and juveniles. Why our results differed so strongly to those obtained by Liu and Chen (2005) cannot be explained. Even if our isolate was not as virulent as their isolate, a low control should have been achieved. Perhaps throughout the duration of the trial conditions briefly changed, which was not registered e. g. pots became too wet or dried out strongly for a short period of time resulting in an ineffectiveness of the fungus in every formulation tested. The trial ought to be repeated with special care to the conditions of the pots. The use of crystal quartz gravel[®], as a carrier for the fungus, has been reported for the first time in this trial. For this reason comparative literature was not found to underline results. Unfortunately, in our studies, results indicated a lack of control. Further studies are needed to check whether, a lack of control was based on a poor formulation or abiotic factors that might have inhibited *H. rhossiliensis*. In our trial, besides crystal quartz gravel[®], wheat seeds colonized with *H. rhossiliensis* were used to introduce *H. rhossiliensis* into the soil. Seeds of wheat, rye and rice have been used to introduce other endoparasitic fungi into the soil (Kerry, 1984). Jaffee and Zehr (1985) also implied that wheat seeds were a good substrate for *H. rhossiliensis* under axenic conditions. They found that *H. rhossiliensis* could be recovered after placing colonized wheat seeds in sterile soil for two weeks but not if seeds were incubated in non-sterile soil. The duration of the trial was 8 weeks in total, during this period, soil which had been sterilized, eventually became insterile through environmental factors. A solid culture of *H. rhossiliensis* on wheat seeds became infested with mold, even though the entire seed had been colonized by *H. rhossiliensis*. Wheat seeds extracted from soil after harvest were also contaminated by mold. One hundred ml biotests conducted with solid culture of *H. rhossiliensis* on wheat seeds showed a good efficacy in suppression of root penetration but as the trials did not last longer than 3 to 4 weeks, perhaps the degree of the contamination was less severe. Nonetheless, a solid culture of *H. rhossiliensis* on wheat seeds for commercial use of *H. rhossiliensis* is inappropriate. A solid culture of *H. rhossiliensis* on corn grits was also applied in the 300 ml pot trial. An application of corn grits without fungus had a better control than the corn grit culture of *H. rhossiliensis* and suppressed nematode numbers by 75%, although the difference was not statistically significant to the control. Liu and Chen (2001) and Chen and Liu (2005) have reported similar results and discussed possible reasons for suppression. They suggested that suppression was caused by toxic effects of the decomposed compounds of corn grits since soil organisms were initially eliminated by the heat-treatment and recolonisation by parasites and predators of nematodes was unlikely. It is a common assumption that an organic carrier for biological agents may help the organism to establish in soil (Stirling, 1991). On the other hand this organic carrier cannot protect the fungus from antagonism. Perhaps the performance of the formulation could be improved if a formulation would better resemble the parasitized nematode. Jaffee (2000) reported that *H. rhossiliensis* alginate pellet formulation was more sensitive to biotic inhibition than the natural inoculum of parasitized nematodes. He also suggested alginate capsules containing *H. rhossiliensis* would perform better if they resembled parasitized nematodes. The most important differences between the two is in the presence of a cuticle or other barrier (the capsule has none) and in the condition of the assimilative hyphae (those in capsules are fragmented in a blender and those in nematodes are intact). As the organic carrier or added nutrients are extremely vulnerable to attack or consumption by aggressive and competitive saprophytes, it is probably most sensible to apply either a fungal suspension of vegetative assimilative hyphae that have unfortunately been blended and are not protected by a cuticle or otherwise parasitized nematodes. The advantage of the application of a fungal suspension is its better distribution throughout the soil obtained by such an application. As parasitized nematodes are extremely small, a very large amount is needed for a sufficient distribution throughout the soil.

125

In general, an effective control of *H. schachtii* by a treatment of liquid and solid formulations of *H. rhossiliensis* was not achieved. Perhaps a protection of the seedling is possible in the first few weeks after germination but the final nematode population density remains unaffected by such an application. Natural infestations of *Hirsutella rhossiliensis* have been reported to suppress *H. schachtii* (Müller, 1985), the fungus was formulated to improve and secure its effectiveness against plant-parasitic nematodes and make it ready for commercial use. Unfortunately, finding a formulation better suitable for the fungus than its natural inoculum has been impossible until now.

9.5 Summary

The efficacy of the endoparasitic fungus *Hirsutella rhossiliensis* was evaluated using four different liquid formulations and three solid formulations against *Heterodera schachtii*. The four liquid formulations as well as a solid culture of *H. rhossiliensis* on wheat seeds achieved a reduction of nematode invasion into the roots of sugar beet seedlings, thus improving plant health. On the contrary, regardless of the liquid or solid formulation applied, nematode reproduction could not be decreased. As an effective biocontrol agent aims to reduce nematode numbers in soil, the formulations tested do not qualify as potential biocontrol agents.

10 Literature

AMIN, A. W. 2000. Efficacy of *Arthrobotrys oligospora, Hirsutella rhossiliensis, Paecilomyces lilacinus* and *Pasteuria penetrans* as potential biocontrol agents against *Meloidogyne incognita* on tomato, Pakistan J. Nematol. 18, 29-33.

BARRON, G. L. 1975. Detachable adhesive knobs in *Dactylaria candida*. Trans. Brit. Mycol. Soc. 65, 311-312.

BAST, E. 2001. Kultivierung von Mikroorganismen. Pp. 47-135 in Bast, E., Mikrobiologische Methoden, 2. Aufl., Spektrum Akademischer Verlag GmbH, Heidelberg, Berlin.

BIRD, A. F. 1959. The attractiveness of roots to plant parasitic nematodes, *Meloidogyne javanica* and *M. hapla*. Nematologica 4, 322-335.

BRUCKNER, S. 2002. Influence of synthetic pesticides on the effectiveness of the *Coniothyrium minitans* based Bio-fungicide Contans® WG. Bulletin OILB/SROP 25 (10), 97.

BURROWS, P. R., Kerry, B. R. and Perry, R. N. 1994. Brief reviews: New approaches to plant-parasitic nematodes. J. Zool., Lond. 232, 341-346.

BYRD, D. W., Jr., Kirkpatrick, T. and Barker, K. R. 1983. An improved technique for clearing and staining plant tissues for detection of nematodes. J. Nematol. 15, 142-143.

CABANILLAS, E., Barker, K. R. and Daykin, M. E. 1988. Histology of the interactions of *Paecilomyces lilacinus* with *Meloidogyne incognita* on tomato. J. Nematol. 20 (3), 362-365.

CARNEIRO, R. M. D. G., Almeida, M. R. A., Quйнйihervй. 2000. Enzyme phenotypes of *Meloidogyne* Spp. Populations. Nematology, 2(6), 645-654.

CASTRO, C. E., Belser, N. O., McKinney, H. E. and Thomason, I. J. 1990. Strong repellency of the root knot nematode, *Meloidogyne incognita* by specific inorganic ions. J. Chem. Ecol. 16, 1199-1205.

CAVENESS, F. E. and Jensen, H. J. 1955. Modification of the centrifugal-flotation technique for the isolation and concentration of nematodes and their eggs from soil and plant tissue. Proc. Helmin. Soc. Wash. 22, 87-89.

CAYROL, J. C., Castet, R. and Samson, R. A. 1986. Comparative activity of different *Hirsutella* species towards three plant parasitic nematodes. Rev. Nématol. 9, 412-414.

CAYROL, J. C. and Frankowski, J. P. 1986. Influence of the number of parasitizing conidia of *Hirsutella rhossiliensis* on the mortality of *Ditylenchus dipsaci*. Rev. Nématol. 9, 411-412.

CHEN, S. 1997. Infection of *Heterodera glycines* by *Hirsutella rhossiliensis* in a Minnesota soybean field, J. Nematol. 29, 573 (abstract).

CHEN, S. 2004. Biological control of nematodes with *Hirsutella minnesotensis*. United States patent 6749848, http://www.freepatentsonline.com/6749848.html.

CHEN, S. Y. and Liu, X. Z. 2005. Control of the soybean cyst nematode by the fungi *Hirsutella rhossiliensis* and *Hirsutella minnesotensis* in greenhouse studies. Biol. Contr. 32, 208-219.

CHEN, S. Y. and Reese, C. D. 1999. Parasitism of the nematode *Heterodera glycines* by the fungus *Hirsutella rhossiliensis* as influenced by crop sequence. J. Nematol. 31, 437-444.

COOKE, D. A. 1984. The relationship between numbers of *Heterodera schachtii* and sugar beet yields on mineral soil. 1978-81. Ann. of Appl. Biol. 106, 111-120.

COOKE, D. 1993. Nematode parasites of sugar beet. Pp. 133-170 in K. Evans, D. L. Trudgill, and J. M. Webster, eds. Plant parasitic nematodes in temperate agriculture. CAB International Wallingford, UK.

DE LEIJ, F. A. A. M., Kerry, B. R., and Dennehy, J. A. 1993. *Verticillium chlamydosporium* as a biological control agent for *Meloidogyne incognita* and *M. hapla* in pot and micro-plot tests. Nematologica 39, 115-126.

DETHIER, V. G. 1947. Chemical insect attractants and repellents. Philadelphia, The Blakison Company, 289 pp.

DIEZ, J. A. and Dusenbery, D. B. 1989. Repellent of root-knot nematodes from exudate of host roots. J. Chem. Ecol. 15, 2445-2555.

DUSENBERY, D. B. 1987. Prospects for exploiting sensory stimuli in nematode control. Pp. 131-135, in J. A. Veech and D. W. Dickson, eds. Vistas in Nematology. Society of Nematology, Hyattsville, Maryland.

DIJKSTERHUIS, J., Veenhuis, M., and Harder, W. 1990. Ultrastructural study of adhesion and initial stages of infection of nematodes by conidia of *Drechmeria coniospora*. Mycol. Res. 94, 1-8.

DOWE, Asmus.1987. Räuberische Pilze und andere pilzliche Nematodenfeinde. A. Ziemsen Verlag, Wittenberg.

DRECHSLER, C. 1937. Some hyphomycetes that prey on free-living terricolous nematodes. Mycologia 29, 447-552.

EAYRE, C. G., Jaffee, B. A., Zehr, E. I., 1987. Suppression of *Criconemella xenoplax* by the nematophagous fungus *Hirsutella rhossiliensis*. Plant Dis. 71, 832-834.

FRANKLIN, M. T. 1951. The cyst-forming species of *Heterodera*. Commonwealth Agricultural Bureaux, Farnham Royal.UK.

GLOVATSKAYA, M.P. 1971. Effect of *Heterodera* infection on photosynthesis of sugar beet in Parazity zhivotnykh I rastenii. Kishinev: Izdatel'stvo 'Shtiintsa'7, 124-130.

GOFFART, H. 1958. Methoden zur Bodenuntersuchung auf zystenbildende Nematoden. Nachrichtenbl. Deut. Pflanzenschutzd. 10, 49-53.

GRECO, N., Brandonisio, A. and de Marinis, G. 1982. Tolerance limit of the sugarbeet to *H. schactii*. J. Nematol. 14, 199-202.

GRIFFIN, G. D. 1981. The relationship of plant age, soil temperature and population density of *Heterodera schachtii* on the growth of sugar beet. J. Nematol. 13, 184-190.

GUTBERLET, V. 2000. Untersuchungen zur Eignung der Verkapselung des nematophagen Pilzes *Hirsutella rhossiliensis* für die biologische Bekämpfung von *Heterodera schachtii* und *Meloidogyne incognita* unter Verwendung nachwachsender Rohstoffe. Dissertation Univ. Bonn.

HALLMANN, J. 1994. Einfluß und Bedeutung endophytischer Pilze für die biologische Bekämpfung des Wurzelgallennematoden *Meloidogyne incognita* an Tomate. Dissertation University of Bonn.

HALLMANN, J., Rodríguez-Kábana, R. and Kloepper, J. W. 1999. Chitin-mediated changes in bacterial communities of the soil, rhizosphere and within roots of cotton in relation to nematode control. Soil Biol. Biochem. 31, 551-560.

HUSSEY, N. W., Wyatt, I. J. and Hesling, J. J. 1969. Trials on the use of methyl-bromide as a substitute for the »cook-out« procedure in mushroom houses. Report of Glasshouse Research Institute for 1961, 63-64.

HUSSEY, R.S. and Barker, K.R. 1973. A comparison of methods of collecting inocula of *Meloidogyne* spp. including a new technique. Plant Dis. Rep. 57, 1025-1028.

IRVING, F., and Kerry, B.R. 1986. Variation between strains of the nematophagous fungus, *Verticillium chlamydosporium* Goddard. II. Factors affecting parasitism of cyst nematode eggs. Nematologica 32, 474-485.

ISEMER, C. 1996. Verkapselung von *Hirsutella rhossiliensis* zur biologischen Bekämpfung des Zuckerrübennematoden *Heterodera schachtii*. Diplomarbeit (thesis) Univ. Oldenburg.

JAFFE, H., Huettel, R. N.,Demilo, A. B., Hayes, D. K. and Rebois, R. V. 1989. Isolation and identification of a compound from soybean cyst nematode, *Heterodera glycines*, with sex pheromone activity. J. Chem. Ecol. 15, 2031-2043.

JAFFEE, B. A. 1992. Population biology and biological control of nematodes. Can. J. Microbiol. 38, 359-364.

JAFFEE, B. A. 1999. Enchytraeids and nematophagous fungi in tomato fields and vineyards. Phytopathology 89 (5), 398-406.

JAFFEE, B. A. 2000. Augmentation of soil with the nematophagous fungi *Hirsutella rhossiliensis* and *Arthrobotrys haptotyla*. Phytopathology 90 (5), 498-504.

JAFFEE, B. A., Gaspard, J. T., and Ferris, H. 1989. Density-dependent parasitism of the soil-borne nematode *Criconemella xenoplax* by *Hirsutella rhossiliensis*. Microb. Ecol. 17, 193-200.

JAFFEE, B. A. and McInnis, T. M. 1990. Effects of Carbendazim and the nematophagous fungus *Hirsutella rhossiliensis* and the ring nematode. J. Nematol. 22 (3), 418-419.

JAFFEE, B. A. and Muldoon, A. E. 1989. Suppression of cyst nematode by natural infestation of a nematophagous fungus. J. Nematol. 21, 505-510.

JAFFEE, B. A. and Muldoon, A. E. 1995. Suppression of the root-knot nematode *Meloidogyne javanica* by alginate pellets containing the nematophagous fungi *Hirsutella rhossiliensis*, *Monacrosporium cionopagum* and *M. ellipsosporum*. Biocontr. Sci. Technol. 7, 203-217.

JAFFEE, B. A. and Muldoon, A. E. 1997. Suppression of the root-knot nematode *Meloidogyne javanica* by alginate pellets containing the nematophagous fungi *Hirsutella rhossiliensis*, *Monacrosporium cionopagum* and *M. ellipsosporum*. Biocontrol Sci. Technol. 7, 203-217.

JAFFEE, B. A., Muldoon, A. E., Anderson, C. E. and Westerdahl, B. B. 1991. Detection of the nematophagous fungus *Hirsutella rhossiliensis* in California sugar beet fields. Biol. Contr. 1, 63-67.

JAFFEE, B. A., Muldoon, A. E., and Didden, W. A. M. 1997. Suppression of nematophagous fungi by enchytraeid worms: A field enclosure experiment. Oecologia 112, 412-423.

JAFFEE, B. A. and Muldoon, A. E., Phillips, R. and Mangel, M. 1990. Rates of spore transmission, mortality, and production for the nematophagous fungus *Hirsutella rhossiliensis*. Phytopathology 80, 1083-1088.

JAFFEE, B. A., Muldoon, A. E. and Westerdahl, B. B. 1996. Failure of mycelial formulation of the nematophagous fungus *Hirsutella rhossiliensis* to suppress the nematode *Heterodera schachtii*. Biol Contr. 6, 340-346.

JAFFEE, B. A. Phillips, R., Muldoon, A. E. and Mangel, M. 1992. Density-dependent host-pathogen dynamics in soil microcosms. Ecology 73, 495-506.

JAFFEE, B. A. and Zasoski, R. J. 2001. Soil pH and the activity of a pelletized nematophagous fungus. Phytopathology 91, 324-330.

JAFFEE, B. A. and Zehr, E. I. 1982. Parasitism of the nematode *Criconemella xenoplax* by the fungus *Hirsutella rhossiliensis*. Phytopathology 72, 1378-1381.

JAFFEE, B. A., and Zehr, E. I. 1983. Sporulation of the fungus *Hirsutella rhossiliensis* from the nematode *Criconemella xenoplax*. Plant Dis. 67,1265-1267.

JAFFEE, B. A. and Zehr, E. I. 1985. Parasitic and saprophytic abilities of the nematode attacking fungus *Hirsutella rhossiliensis*. J. Nematol. 17 (3), 341-345.

JENKINS, W.R. 1964. A rapid centrifugal-flotation technique for separating nematodes from soil. Plant Dis. Rep. 48, 692.

JONZ, M. G., Riga, E., Mercier, A. J. and Potter, J. W. 2004. Partial isolation of a water soluble pheromone from the sugar beet cyst nematode, *Heterodera schachtii*, using novel bioassay. Nematology 3 (1), 55-65.

KAYA, H. K. 1985. Entomogenous nematodes for insect control in IPM systems. Pp. 283-302 in M. A. Hoy and D. C. Herzog, eds. Biological Control in Agricultural IPM Systems. Academic Press, Orlando, Florida.

KERRY, B. R. 1984. Nematophagous fungi and the regulation of nematode populations in soil. Helminthological Abstracts, Series B 53, 1-14.

KERRY, B. R. and Jaffee, B. A. 1997. Fungi as biological control agents for plant parasitic nematodes Pp. 203-218 in D. T. Wicklow and Söderstrom, B. E., eds. in The Mycota IV Environmental and microbial relationship; A comprehensive treatise on fungi as experimental systems for basic and applied research, eds. K. Esser and P.A. Lemke; Springer-Verlag Berlin Heidelberg.

KLINGLER, J. 1963. Die Orientierung von *Ditylenchus dipsaci* in gemessenen künstlichen und biologischen CO_2-Gradienten. Nematologica 9, 185-199.

KLINGLER, J. 1965. On the orientation of plant nematodes and of some other soil animals. Nematologica 11, 4-18.

KNUDSON, G. R., Eschen, D. J., Dandurand, L. M., and Wang, Z. G. 1991. Method to enhance growth and sporulation of pelletized biocontrol fungi. Appl. Environ. Microbiol. 57, 2864-2867.

KRALL, E. L. and Krall, Kh. A. 1978. Revision of the plant nematodes of the family *Heteroderidae* on the basis of the trophic specialization of these parasites and their co-evolution with their host plants. Filogelmintologicheskie issledovaniya Moskow, USSR, »Nauka«, 39-56.

LACKEY, B. A., Jaffee, B. A. and Muldoon, A. E. 1992. Sporulation of the nematophagous fungus *Hirsutella rhossiliensis* from hyphae produced *in vitro* and added to soil. Phytopathology 82, 1326-1330.

LACKEY, B. A., Muldoon, A. E. and Jaffee, B. A. 1993. Alginate pellet formulation of *Hirsutella rhossiliensis* for biological control of plant-parasitic nematodes. Biological Control 3, 155-160.

LACKEY, B. A., Muldoon, A. E. and Jaffee, B. A. 1994. Effect of nematode inoculum on suppression of root-knot and cyst nematodes by the nematophagous fungus *Hirsutella rhossiliensis*. Phytopathology 84, 415-420.

LACKEY, B. A., Jaffee, B. A., and Muldoon, A. E. 1992. Sporulation of the nematophagous fungus *Hirsutella rhossiliensis* from hyphae produced *in vitro* and added to soil. Phytopathology 82, 1326-1330.

LACKEY, B. A., Jaffee, B. A., and Muldoon, A. E. 1993. Alginate pellet formulation of *H. rhossiliensis* of *Hirsutella rhossiliensis* for biological control of plant-parasitic nematodes. Biol. Control 3, 155-160.

LEWIS, J. A., and Papavizas, G. C. 1984. A new approach to stimulate population proliferation of *Trichoderma spp.* and other potential biocontrol fungi introduced into natural soils. Phytopathology 74, 1240-1244.

LIEBMAN, J. A., and Epstein, L. 1992. Activity of fungistatic compounds from soil. Phytopathology 82,147-153.

LIU, X. Z. and Chen, S. Y. 2000. Parasitism of *Heterodera glycines* by *Hirsutella* spp. in Minnesota soybean fields. Biol. Contr. 19, 161-166.

LIU, X. Z and Chen, S. Y. 2001. Screening isolates of *Hirsutella* species for biocontrol of *Heterodera glycines*. Biocontrol Sci. Technol. 11, 151-160.

LIU, S. F. and Chen, S. Y. 2002. Effect of pH on growth and sporulation of *Hirsutella minnesotensis* and *Hirsutella rhossiliensis in vitro*. Phytopathology 92, S48.

LIU, S. and Chen, S. 2005. Efficacy of the fungi *Hirsutella minnesotensis* and *H. rhossiliensis* from liquid culture for control of the soybean cyst nematode *Heterodera glycines*. J. Nematol. 7(1), 149-157.

LOCKWOOD, J. L. 1988. Evolution of concepts associated with soil borne plant pathogens. Annu. Rev. Phytopathol. 26, 93-121.

MCCOY, C. W. 1990. Entomogenous fungi as microbial pesticides. Pp. 139-159 in R. R. Baker and P. E. Dunn, eds. New directions in biological control: Alternatives for suppressing agricultural pests and diseases. New York: Alan R. Liss.

MCINNIS T. M & Jaffee, B. A. 1989. An assay for *Hirsutella rhossiliensis* spores and the importance of phialides for nematode inoculation. J. Nematol. 21, 229-235.

MEYER, S. L. F., Johnson, G., Dimock, M., Fahey, J. W., and Huettel, R. N. 1997. Field efficacy of *Verticillium lecanii*, sex pheromone, and pheromone analogs as potential management agents for soybean cyst nematode. J. Nematol. 29, 282-288.

MINTER, D. W. and Brady, B. L. 1980. Mononematous species of *Hirsutella*. Trans. Brit. Mycol. Soc. 74, 271-282.

MÜLLER, J. 1980. Ein verbessertes Extraktionsverfahren für *Heterodera schachtii*. Nachrichtenbl. Deut. Pflanzenschutzd. 32, 21-24.

MÜLLER, J. 1982. The influence of fungal parasites on the population dynamics of *Heterodera schachtii*. Nematologica 28, 161 (abstract).

MÜLLER, J. 1985. Aussichten des integrierten Pflanzenschutzes bei der Bekämpfung pflanzenparasitärer Nematoden. Gesunde Pflanzen 37, 216-221.

MÜLLER, J. 1992. Detection of pathotypes by assessing the virulence of *Heterodera schachtii*.

MÜller, J. 1999. The economic importance of *Heterodera schachtii* in Europe. Helminthologia 36 (3), 205-213.

NÚÑEZ-FERNÁNDEZ, M. C. 1992. Scanning and transmission electron microscopy studies on predator-prey relationships between *Hirsutella* spp. (Patouillard) and some nematode genera. Thesis, Faculteit van de Landbouwwetenschappen, Universiteit van Gent.

131

ORTON Williams, K. J. 1975. *Meloidogyne incognita*. Pp. 4. in CIH Descriptions of plant-parasitic nematodes Set 5, No. 62, Commonwealth Institute of Helminthology, St Albans, UK.

PAPAVIZAS, G. C., and Lewis, J. A. 1981. Introduction and augmentation of microbial antagonists for the control of soil-borne plant pathogens. Pp. 305-322 in: Biological Control in Crop Protection BARC Symposium No. 5. G. C. Papavizas, ed. Allanheld & Osmund, Totowa, NJ.

PATEL, A., Isemer, C, Müller, J., Vorlop, K.-D. 1996. Bekämpfung von phytopathogenen Nematoden mit verkapselten nematophagen Pilzen. Mitt. Biol. Bundesanstalt. Land-Forstwirtsch. 321, 436.

PATEL, A. 1998. Verkapselungsverfahren für die biologische Schädlingsbekämpfung und zur Konstruktion von »vegetativem Samen«. Sonderheft Landbauforschung Völkenrode Vol. 188.

PATEL, A., Slaats, B. E., Hallmann, J., Tilcher, R., Beitzen-Heineke and W., Vorlop, K.-D. 2004. Verkapselung von bakteriellen Antagonisten und eines nematophagen Pilzes. Gesunde Pflanzen 57, 30-33.

PLINE, M., and Dusenberry, D. B. 1987. Responses of the plant-parasitic nematode *Meloidogyne incognita* to carbon dioxide determined by video camera-computer tracking. J. Chem. Ecol. 13, 1617-1624.

POINAR, G.O., and Leutenegger, R. 1968. Anatomy of the infective and normal third-stage juveniles of *Neoplectana carpocarpse* Weiser (Steinernematidae: Nematoda). J. Parasitol., 54, 340-350.

POINAR, G. O., and Jansson, H. B. 1986. Infection of Neoaplectana and *Heterorhabditis* (Rhabdita: Nematoda) with the predatory fungi, *Monacrosporium ellipsosporum* and *Arthrobotrys oligospora* (Moniliales: Deuteromycetes). Rev. Nematol., 9, 241-244.

PROT, J. C. 1980. Migration of plant-parasitic nematodes towards plant roots. Rev. Nematol. 3, 305-318.

PROT, J.-C. and Van Gundy, S. D. 1981. Effect of soil texture and the clay component on migration of *Meloidogyne incognita* second-stage juveniles. J. Nematol. 13, 213-217.

RICHARDSON, P. N. and Grewal, P. S. 1993. Nematode pests of glasshouse crops and mushrooms. Pp. 501-544 in K. Evans, D. L. Trudgill, and J. M. Webster, eds. Plant parasitic nematodes in temperate agriculture. CAB International, University Press, Cambridge Wallingford, UK.

ROBINSON, A. F. and Jaffee, B. A. 1996. Repulsion of *Meloidogyne incognita* by alginate pellets containing hyphae of *Monacrosporium cionopagum, M. ellipsosporum*, or *Hirsutella rhossiliensis*. J. Nematol. 28 (2), 133-147.

ROSE, T. 2000. Fermentation und Verkapselung des nematophagen Pilzes *Hirsutella rhossiliensis* in Polyelektrolyt-Hohlkugeln zur biologischen Bekämpfung des Nematoden *Heterodera schachtii*. Dissertation Techn. Univ. Carolo-Wilhelmina Braunschweig.

SCHUSTER, R.-P., and Sikora, R. A. 1992. Influence of different formulations on fungal egg pathogens in alginate granules on biological control of *Globodera pallida*. Fundam. Appl. Nematol. 15, 257-263.

SINGH, R. S. and Sitaramaiah, K. 1994. Pp. 4 in Plant pathogens, The nematodes. International Science Publisher, New York.

SCHMIDT, A. (1871) Über den Rübennematoden. Zeitschrift Ver. Rübenzuckerindustrie Zollver., 22, 67-75.

SCHUSTER, R. P. and Sikora, R. A. 1992. Influence of different formulations of fungal egg pathogens in alginate granules on biological control of *Globodera pallida*. Fundam. Appl. Nematol. 15, 257-263.

SHEPHERD, A. M. 1955. Formation of the infection bulb in *Arthrobotrys oligospora* Fresenius. Nature 175, 475.

STIRLING, G. R. 1991. Biological control of plant-parasitic nematodes: Progress, problems and prospects. Wallingford, UK, CAB International, 282 pp.

STIRLING, G. R. and Smith, L. J. 1998. Field tests of formulated products containing either *Verticillium chlamydosporium* or *Arthrobotrys dactyloides* for biological control of rootknot nematodes. Biol. Contr. 11, 231-239.

STEINER, G. 1925. The problem of host selection and host specialization of certain plant-infesting nemas and its application in the study of nemic pests. Phytopathology 15, 499-534.

STURHAN, D. and Schneider, R. 1980. *Hirsutella heteroderae*, ein neuer nematodenparasitärer Pilz. Phytopath. Z. 99, 105-115.

TEDFORD, E. C., Jaffee, B. A. and Muldoon, A. E. 1992. Effect of soil moisture and texture on transmission of the nematophagous fungus *Hirsutella rhossiliensis* to the cyst and rootknot Nematodes. Phytopathology 82, 1002-1007.

TEDFORD, E. C., Jaffee, B. A. and Muldoon, A. E. 1994. Variability among isolates of the nematophagous fungus *Hirsutella rhossiliensis*. Mycol. Res. 98, 1127-1136.

TEDFORD, E. C., Jaffee, B. A. and Muldoon, A. E. 1995. Suppression of the nematode *Heterodera schachtii* by the fungus *Hirsutella rhossiliensis* as affected fungus population density and nematode movement. Phytopathology 85 (5), 613-617.

TIMPER, P. and Brodie, B. B. 1993. Infection of *Pratylenchus penetrans* by pathogenic fungi. J. Nematol. 25, 297-302.

TIMPER, P. and Brodie, B. B. 1994. Effect of *Hirsutella rhossiliensis* on infection of potato by *Pratylenchus penetrans*. J. Nematol. 26, 304-307.

TIMPER, P. and Kaya, H. K. 1989. Role of the second-stage cuticle of entomogenous nematodes in preventing nfection by nematophagous fungi. J. Invert. Pathol. 54, 314-321.

TIMPER, P., Kaya, H. K. and Jaffee, B. A. 1991. Survival of entomogenous nematodes in soil infested with the nematode-parasitic fungus *Hirsutella rhossiliensis* (Deuteromycotina: Hyphomycetes). Biol. Contr. 1, 42-50.

VELVIS, H. and Kamp, P. 1995. Infection of second stage juveniles of potato cyst nematodes by the nematophagous fungus *Hirsutella rhossiliensis* in Dutch potato fields. Nematologica 41, 617-627.

VELVIS, H. and Kamp, P. 1996. Suppression of potato cyst nematode root penetration by the endoparasitic nematophagous fungus *Hirsutella rhossiliensis*. Eur. J. Plant Pathol. 102, 115-122.

VIAENE, N. M. and Abawi, G. S. 2000. *Hirsutella rhossiliensis* and *Verticillium chlamydosporium* as biocontrol agents of the root-knot nematode *Meloidogyne hapla* on lettuce. J. Nematol. 32, 85-100.

WARD, S. 1973. Chemotaxis by the nematode *Caenorhabditis elegans*: Identification of attractants and analysis of the response by use of mutants. Proc. Natl. Acad. Sci. USA 70, 817-821.

WIESER, W. 1956. The attractiveness of plants to larvae of root-knot nematodes. II. The effect of excised bean, eggplant, and soybean roots on *Meloidogyne hapla* Chitwood. Proc. Helminth. Soc. Wash. 23, 59-64.

WEIßENBORN, G. 1995. Immobilisierung des Pilzes *Hirsutella rhossiliensis* in Kapseln auf Basis von Polyelektrolyten zur Bekämpfung des Zuckerrübennematoden *Heterodera schachtii*. Diplomarbeit (thesis) Techn. Univ. Carolo-Wilhelmina Braunschweig.

WOODWARD, J. E., Walker, N. R., Dillwith, J. W., Zhang, H. and Martin, D. L. 2005. The influence of fungicides on *Arthrobotrys oligospora* in simulated putting green soil. Ann. Appl. Biol. 146, 115-121.

11 Acknowledgements

I would like to acknowledge the debt I owe to Prof. Dr. Richard A. Sikora for providing me with not only a »Diplom-thesis« in his Department of Soil Ecosystems Phytopathology & Nematology but also a doctorate thesis. Although research for each was conducted in a laboratory far from his department, I always felt like a BÖS and enjoyed our excursions to the fullest. The best excursions were our trips to the USA and Bulgaria, where I could experience the luck I had for having a professor that received highest recognition from leading scientists and nematologists worldwide. His knowledge and personality has enriched me.

I would like to thank Dr. Johannes Hallmann for providing the shelter conditions under which my work could take place. Also, I am particularly grateful to him for his constant supervision of my work and for acting as a »substitute Doktorvater« during my stay in Münster at the Federal Biological Research Centre for Agriculture and Forestry. He was also characteristically generous in taking time to review and help me with my research papers and doctorate thesis.

I kindly thank Prof. W. Amelung for giving his valuable time as a second supervisor.

My special thanks go to my project partners Dr. Anant Patel, Wilhelm Beitzen-Heineke, Ulli Bilgeshausen and Dr. Ralf Tilcher who have not only supported me with encouragement but have supplied me with the various capsule formulations and sugar beet pills. Many thanks to Ulli for material support and Anant for spiritual, inspiring support at critical and opportune times. Their friendship and professional collaboration meant a great deal to me.

I am indebted to all my collegues who have assisted me in one way or another, especially Valeska, Agnes, Falko and Mechthild S. for showing me the ropes, Dagmar for putting up with me in her lab for so many years and for becoming such a dear friend and Engelbert, Björn and Jens for the inspiring but also relaxing lunch breaks, which I will miss for sure.

Many thanks also go to all my friends and collegues at BÖS, especially Tesfa, Tam, Amer and Alexander with who I have shared many enjoyable moments and for their encouragement.

I would like to express sincere gratitude to the Agency for Renewable Ressources (FNR) for their financial support for my research.

I am very grateful to my father who was willing to revise this thesis, although he's not a nematologist. Thanks for your determination. Furthermore, I could not thank my parents and siblings Joseph and Angelique enough for their spiritual support.

I am also grateful to Michael Otto for his aid while formatting this thesis.

Finally, I would like to express my sincere gratitude to Matthias Donner for motivating me and supporting me throughout the entire path of my studies, indifferent to my whereabouts. I could not have imagined a better friend, without you I would not be where I am today.